Labyrinth

A Search for the Hidden Meaning of Science

Peter Pesic

The MIT Press
Cambridge, Massachusetts
London, England

First MIT Press paperback edition, 2001

This book was set in Sabon by Achorn Graphic Services, Inc. on the Miles System.

Printed and bound in the United States of America.

Portions of this book appeared originally in the following journals, which have kindly given permission for the appearance of the material here: *Cryptologia, Interpretation* (© 1999), *Isis* (© The History of Science Society, University of Chicago Press 1997, 1999), *Literature and Theology* (© Oxford University Press, 1994).

Permission for use of the figures has been kindly given by the following: Grace K. Babson Collection of the Works of Sir Isaac Newton (figure 9.2); Beinecke Rare Book and Manuscript Library, Yale University (figure 6.2); The Burndy Library, Dibner Institute for the History of Science and Technology (figures 1.1, 1.2, 9.1); William Donahue and the Cambridge University Press (figure 8.4); Warburg Institute (figures 2.1, 6.1, 8.1, Copyright Warburg Institute). I am grateful to William Donahue for his generous help in providing the Kepler figures. I also thank Elizabeth McGrath (Warburg Institute), Tanya Bresinsky and Ben Weiss (Burndy Library), and Clark Elliott (Babson Collection) for their help with the illustrations.

Library of Congress Cataloging-in-Publication Data

Pesic, Peter
Labyrinth : a search for the hidden meaning of science / Peter Pesic.
p. cm.
Includes bibliographical references and index.
ISBN 0-262-16190-7 (hc : alk. paper), 0-262-66126-8 (pb)
1. Science—Philosophy. 2. Science—Methodology. I. Title.
Q175.P3865 2000
501—dc21 99-058366

Dedicated to my parents,
who taught me the meaning of courage

Contents

Prelude

When Albert Einstein was four or five, his father showed him a compass. Watching it, the child felt that there was something deeply hidden behind things. He trembled and grew cold. This was a turning point for him, but he also realized something beyond the merely personal: nature has secrets. This idea is strange, despite its familiarity. "Secrets" imply a mysterious depth behind appearances; it suggests some reason for secrecy and the possibility of discovering nature's hidden laws. But what makes them secret? Is it because they are far beyond human ken? Then it would be futile or presumptuous to try to solve them. Is it because they concern divine matters? Then it might be dangerous to touch them. If they are accessible to human seeking, why should they be so carefully hidden? And if so hidden, how could they ever be disclosed? These questions about the hidden quest of science and its human meaning are the crux of this book.

My inquiry requires viewing familiar things as if they were strange and questionable. This reflects the emergent character of human thought. The original meaning of a concept is not like a geological stratum, buried far beneath the surface, but is still alive and active. Thus the origin of science is not fossilized, dead sediment, but a continuing presence that informs the full reality of living science. This living center touches the concerns of thoughtful people. I do not presume that the reader knows the writings that I will discuss, whether scientific or literary; I have tried to bring them to life anew. Most of all, I want to open up vivid questions that will continue to live in the mind so that, even though a book is a silent companion, this one might begin a true conversation.

There have long been two contrasting attitudes toward the intelligibility of the world. Some ancient thinkers maintained that that the world is open to human intelligence. Others held there are hidden depths that lie beyond human sensibility. The enigmatic Heraclitus wrote that "Nature loves to hide," and added that "unless you expect the unexpected you will never find it, for it is hard to discover and hard to attain." In contrast, Aristotle discerned a harmony between our sensibility and the world in which we live. The Aristotelian scientist is a patient and observant onlooker, not a meddling experimenter. Aristotle's respectful insights into nature disclose beauty rather than power. He directs our gaze to the bright stage of nature, where everything will appear in due course.

Though modern science owes a great deal to Aristotle, it also seeks hidden powers. Several traditions held those secrets in reserve. Though acting in history, the Biblical God is "a hidden God indeed," hallowing the mysterious depths. The scientific quest for those depths has a special intensity. If one can speak of a religious quality to science, it emerges in the ardent search for hidden connections and may not be bound by established traditions. Though the scientists in this book are variously Christian, Jewish, or skeptic, they all seek an ultimate encounter with the heart of things. Their exhilaration and desolation mark the stages of a spiritual journey.

That journey goes beyond the bounds of received wisdom, even of the old traditions of magic power. Various hermetic writings alluded to powerful secrets hidden from all but the elect. Presuming them to be very ancient, the curious were fascinated by these mysterious promises, which included alchemy and other kinds of esoteric magic. Many collections of "secrets of nature" promised access to those powers for adepts and their princely patrons. Besides dark accounts of occult power, these "secrets" included many practical recipes and technical skills previously guarded as craft mysteries. Though fascinated by these promises and disclosures, the new science turned away from the traditional pursuit of the occult, which repeated old arcana rather than seeking fresh understanding. Nature seemed a labyrinth requiring new heroes capable of finding their way through those unknown, intricate passages.

I will offer a guiding thread that has three main strands. In musical terms, this book is a triple fugue, an interweaving of three distinct but finally interrelated themes concerning the character of the scientific enterprise and the deep effects it has on human character. The first theme is the arduous struggle between the scientist and nature, a heroic ordeal that tests both of them. The second is the effect that struggle has on the character of the scientist, in particular a kind of purification that affects the wellsprings of human desire. I will use the Greek word eros to speak about the whole range of ardent desires, not restrict it to sexuality in the narrow sense usually conveyed by terms like "erotic." This word allows us to think more clearly about what we most desire, and about the nature of our desires. My third theme is a crucial innovation that emerges from these purified desires, symbolic mathematics, and its remarkable relation to the world it symbolizes.

The first three parts of the book introduce these themes in detail, and also begin to show how they intertwine. These opening parts take place in the Renaissance, at particular points of origin central for my themes and their later unfolding. They concern three seminal figures: William Gilbert, a physician who began the scientific study of magnetism; François Viète, a French counselor and codebreaker who played a crucial role in the foundation of modern symbolic mathematics; and especially Francis Bacon, Lord Chancellor of England and the visionary who anticipated the shape of modern science. Each of them grasped a crucial insight at a formative stage, and thus can reveal an emergent theme with a clarity that tends to get lost in later developments. Gilbert emphasized the importance of experience and experiment, but did not consider the possibilities of mathematical theory. Viète, in contrast, was not an experimental scientist, but pioneered modern symbolic mathematics. Their separate work indicates the streams of experiment and theory that were to converge in modern science.

Although Bacon did not perform significant experiments or formulate great theories, I will give him center stage. He foresaw a new engagement with nature, and was the trumpeter who summoned others to battle. As herald, he sounded crucial messages concerning what science should be. He offered the earliest and most influential justification of the new enterprise, as he conceived it. Since he was describing something not yet formed, he used a rich variety of language to evoke his new vision of science and especially turned to ancient myths as a storehouse of potent images that he reinterpreted to illustrate his meaning. Reaching into antiquity, Bacon found depths that foreshadow the scientific quest into the depths of Nature. Reaching toward the future, he used the ancient myth of Atlantis to evoke a new world. His writings are the mirror in which the enigmatic birth of modern science may best be seen.

At first Bacon was idolized. Jean-Jacques Rousseau thought him perhaps the greatest of all philosophers; Thomas Jefferson considered him one of the three greatest men who ever lived; Charles Darwin claimed to be a faithful practitioner of Baconian method. Nevertheless, others not only questioned Bacon's importance, but also impugned his integrity. Thomas Macaulay mixed high praise of Bacon's intellectual achievements with damning censure of what he considered Bacon's political and personal corruption. Various invectives accused Bacon of personal coldness, professional cruelty, and prostitute ambition; these accusations often rely on old canards and misrepresented evidence. The vilification of Bacon's character aided the dismissal of his vision as finally irrelevant to modern science, despite the testimony of his successors. I have tried not to be swayed either by such unjust attacks or by a compensatory desire to avenge Bacon by idolizing him once more.

In truth, these intense and surprising tides of adoration and revulsion indicate not only how important Bacon was, but also how disturbing he continues to be. Perhaps the key is the way he strikes some readers as cold, despite biographical evidence of his devoted friendships and personal loyalties, not to speak of his overarching philanthropic projects. Bacon casts a cold eye on political realities and on the rest of nature, including us. He challenges us to overturn the idols that rule our minds and keep them from the arduous quest for truth. As I will discuss, he scrutinizes our inmost desires, aiming to cool their normal course and warm them toward a new eros of scientific discovery. He intends to disturb our complacent opinions, yet his detachment dares us to read him coolly. If he provokes us, he may also test and help us, for he intends to do us good. Eventually we must confront him, since we live in *his* world now: ours is the age of Bacon.

Accordingly, the full implications of his insights must be sought in modern times and in its master scientists. The final part of the book concerns the encounters of Kepler, Newton, and Einstein with the depths of Nature. In their works, my three themes are densely intertwined. As befits someone groping through a maze, I do not feel restricted in the way I search for the center. The word *clue* originally meant the thread guiding one through a maze, and I will present the clues as I find them, whether in Gilbert's account of magnetism or Bacon's retold fables or Viète's codebreaking. I will not present a continuous history or offer an overview of scientific theories, although I will recount some elements of history and of scientific theory along the way. So, this book is not a work either of the history of science or of the philosophy of science in their usual modern senses. These scholarly disciplines are recent subdivisions of a larger, more inclusive whole that was originally called *philosophy,* the "love of wisdom." This meant far more than the technical and restricted field that modern philosophy has become. I will read scientific works as if they were works of literature, attending to nuance and tone as much as to surface meaning, trying to be faithful to them. These are works of thought and feeling that call for the evocation of their largest questions; they resist the imposition of false mysticism or facile simplification. Like great works of literature, great scientific writings have a compelling integrity and imaginative force that call for sensitive reading. In this spirit, I try to find the living center of human concern as it emerges in the scientific endeavor. Though I cannot claim to be a philosopher in the deepest sense, I invite the reader to a philosophical quest.

My approach has something of the "outsider" as well as of the "insider," and in this respect also differs from other accounts that restrict themselves to one view or the other. Both perspectives are helpful and important. The outsider does not take science for granted, and is able to see what is strange and remarkable more readily than someone long habituated to scientific ways. The insider is more familiar with the lived experience of science, through intimacy and (one hopes) through love. What for the insider is a compelling encounter with reality may seem to the outsider merely a social convention, a difference of cultures, a mask of social dominance. It is hard to imagine being wholly satisfied with a visitor's report on the life of a tribe, however sympathetic and perceptive the visitor might be. One wants, if possible, to hear the native voice clearly and freshly. To do this requires a search for origins, for living sources.

We have reached the point of departure. Here begins the labyrinth; its winding corridors mirror the intricate turnings of the world and of the seeker. Even beyond matters of science, the quest into the depths of nature may raise the deepest human questions.

I

The Night Watch

1

The Hard Masters

For Einstein, the compass pointed to something hidden behind things; the needle also indicated his own special direction in life. His story can also be read as pointing backward in time to the origin of the modern study of magnetism, to William Gilbert, physician to Queen Elizabeth I. Gilbert's book *On the Magnet* (1600) is a revealing document both for his insights into magnetism and the way he reaches them. He was among the first to test ideas about nature, using experiment in the modern sense; he expresses the novel character of experimental knowledge.

Although the Greeks and Romans had lodestones (naturally magnetic rocks), they did not realize they could be used to indicate north and south; it was the Chinese who first used them to find directions at sea. Thus the most fundamental facts about magnetism were long unknown in the West. Gilbert appropriates much of his lore from mariners and practical men, and he scorns those who merely repeat bookish opinions without using the light of experience, for experiment opens a new path toward reliable knowledge. Gilbert considers that magnetism rests on "hidden things which have no name and that never come into notice," yet which are not dark and obscure, as the alchemists had thought. To replace their mystifying terminology, Gilbert uses plain words and calls for a new style of philosophizing adequate to fathom the hidden depths of the Earth. There, he finds magnetic stones that contain the potency of the Earth's inward substance. Gilbert's novel insight is that the Earth is a great lodestone. The compass seeks the Earth's pole because the compass needle shares the magnetic life of its great parent, "Earth, the mother of all." Aristotle taught that earth is static and inert, "a dead weight planted in the center of the universe at equal distance everywhere from the heavens." Gilbert holds that Earth has living potencies that reach down to its inmost depths.

To prove these assertions, Gilbert fashions out of lodestone a small round image of the Earth, which he calls a terrella or "little Earth." He shows that the terrella will align a tiny compass or versorium put on it, in imitation of a compass on the Earth. Gilbert thus creates a model of the Earth and views it with the eye that God casts over the world. In this microcosm Gilbert sees a parallel between the behavior of versorium and compass that confirms the analogy between the terrella and the whole Earth.

Gilbert uses this model to investigate the variations from true north that mariners had long known. Large continental land masses, composed of magnetic rock, tug on nearby compasses and cause them to deviate from geographical north. Gilbert demonstrates this by sculpting such masses in the microcosm of the terrella (figure 1.1). He even uses the terrella to think about the motions of the Earth. Gilbert admires Copernicus and ridicules those who ascribe a daily rotation to the immense sphere of the fixed stars, rather than to the Earth. He argues that as the little terrella rotates in order to align itself with the Earth, so too must the Earth rotate. Thus Gilbert ascribes the

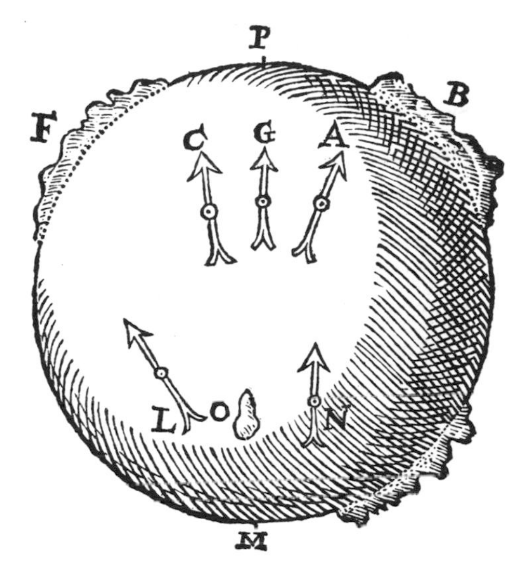

Figure 1.1
Gilbert's terrella, showing miniature "continents" sculpted on it. A miniature compass (vesorium) at G will point to the true north pole (P), while one at A will deviate toward the nearby "continent" at B.

daily rotation of the earth to *magnetic* causes. This conjecture will be important for Kepler, although later investigations did not bear it out.

Gilbert's experiments led him to still more intimate engagement with earth and metal. He shows considerable practical knowledge of the arduous processes that treat iron ore, still earthy in appearance, with intense fire. He also knows of the process of tempering, in which steel is forged by plunging red-hot iron into a quenching water bath; the ordeal of water augments the ordeal of fire, changing raw ore into shining metal, strengthening it through the stress of contradictory extremes. His quest to confront the mystery of magnetic alignment involved arduous trial and would have tested his "temper" as well. He gives a firsthand report of descending into a mine, digging a lodestone out of its mother lode, and checking its orientation. This experience confirms his view that the Earth is a great lodestone since it can imprint its direction on its child, the terrella. He also verifies this by reliving the experience in the little world of the terrella, carving out a miniature mine and extracting an even more diminutive lodestone, a sort of grandchild of Earth. The exact agreement of the two experiences could only hold if terra and terrella are essentially alike.

Despite these searching trials, Gilbert is troubled by the notion that nature's secrets could be wrested from her by brute force. The alchemists held that transmutations require violent incursions into matter, especially through fire. In alchemical writings the metal speaks from the fiery crucible: I am sacrificed, I suffer intolerable violence, I am dismembered, and at last I am transformed. On the contrary, Gilbert notes that a lodestone is sensitive to excessive heat; it "loses its native and inborn powers of attracting, and all other magnetic properties, if left long in fire."

Though fire is a crucial factor in the forging and shaping of metals, it is also a violent intruder in the native magnetic life of the ore. Gilbert discusses "those hard masters of the metals, who in many various processes put them to the torture." Gilbert thinks of the Earth as a mother; in the fiery crucible, iron is tested and verified to be "more truly the child of Earth than any other metal," for it endures the hottest furnaces. Common iron prevails, more earthy than the "noble" metals gold and silver, which cannot withstand the ordeal of fire. Though forcibly taken from its mother, iron emerges victorious, not destroyed but purified. It is the true child of Earth through tested worth, not just by right of birth.

Figure 1.2
Gilbert's illustration of iron being magnetized as it cools.

Yet Gilbert needed to become one of those "hard masters" who remove the child from the mother lode to test and forge it. By recreating the experience in the microcosm of the terrella, Gilbert was able to see it from the standpoint of the Maker and to behold the coherence of the whole: "All the inner parts of the Earth are in union and act in harmony, and produce direction to north and south." By sharing the experience of the miners and the heat of the smithy, Gilbert himself underwent a kind of test, for originally *experiment* meant *experience.* He gave a detailed and realistic picture of the Earth magnetizing iron as it cools (figure 1.2). Furthermore, certain iron ores that are *not* magnetic in their native state can be made magnetic if heated in a moderate fire. This is surprising because fire, applied moderately and in due measure, apparently can induce magnetic behavior in an otherwise indifferent rock. There is some delicate interconnection that moderate fire promotes and extreme fire disrupts.

Gilbert shared the ancient belief that the mysterious self-movement of magnets indicated some kind of soul. As a physician, he was aware of the proper degrees of living heat in the human body and the dangers of fever or chill. He judges that the iron that has been heated red-hot "is not iron, but something lying outside its own nature, until it returns to itself." By deforming the body, fire may disrupt the magnetic soul. When lodestones are burned, sometimes a sulphurous or foul-smelling vapor is exhaled from "the foul bed or matrix in which the lodestone is produced," as if the mother lode were an earthy womb. Yet these material emanations cannot explain magnetic power, which acts even *through* intermediate layers of glass or gold or stone.

Magnetism depends on form. A white-hot iron "has a confused, disordered form, and therefore is not attracted by a lodestone, and even loses its power of attracting, however acquired." If the form of the magnet is distorted by excessive heat, that change is potentially reversible; white-hot iron has been *confused,* not transformed. He compares the iron to the soul, whose powers persist even in a burned body. The form is a kind of soul in which magnetism can be sheltered against violence.

Gilbert chooses the word *coition,* with all its amorous associations, to characterize this magnetic soulfulness. This erotic language goes back to antiquity; "iron is drawn by the lodestone as the bride to the embraces of her spouse." Coition implies a *mutual* coming together, rather than the one-sided "rush toward union which commonly is called attraction." The magnet is a child of Earth; magnetism is the manifestation of their familial bond, emergent from both together, not from either separately. He invokes vivid images of magnets growing in the earth "just as in the warm uterus the seed or the embryo grows"; lodestone and iron "come into being and are generated in the same matrix, in one bed, like twins." Gilbert rejects the astrological explanation that ascribes the generation of metals to the influence of the planets. The metals emerge in a warm love-bed, deep inside the womb of Earth, as with "the blood and the semen in the generation of animals." Thus the erotic life of the Earth is richly fruitful; she shares her magnetism with her children as the sign of their bond.

The life of magnetic bodies seems to him different from that of animals; used as a medicine, lodestone "often is injurious and fatal in its effects." The magnet acts by soul, for it is not "fed" by gross matter. It needs the companionship of iron filings as a kind of cradle that imitates the maternal care of Earth; otherwise the lodestone might age and decline, but not irreversibly. Like a soul, the stone can experience rebirth through the proper application of heat and contact with another magnet.

Nonetheless, Gilbert does not for that reason look with indifference on violence. He rejects the notion of *attraction,* for it suggests that "force seems to be brought in and a tyrannical violence rules." His invocation of love informs his rejection of war. Gilbert speaks with the greatest contempt—ironically, with violent invective—against the opinion that there is a "battle of forces" in the magnetic encounter. His opinions echo those of his royal patient; Queen Elizabeth was adamantly opposed to war as a dangerous drain on the state. Though she acted decisively against the Spanish armada, she was swift to disperse her forces once the threat was repulsed. The horror of sedition was widely shared by her subjects, ever fearful of insurrection or civil war. Gilbert hates the ancient earth-centered cosmology because it embodies "a universal force, an unending despotism, in the governance of the stars, and a hateful tyranny" that forces the fixed stars to circle the Earth.

His violent antipathy stems from his paradoxical peacefulness. Gilbert places magnetic coition on the highest levels of amity and peacefulness, along with "electric coition," his name for static electricity. His tone softens as he describes how the delicate effluvia that emerge from electrically charged substances "lay hold of the bodies which they unite, enfold them, as it were, in their arms, and bring them into union with the electrics." By rejecting war and force Gilbert turns instead toward metaphors of peaceful and legitimate rule; in the magnetic poles "this mighty power has here its chief excellency; here is its throne, so to speak." Here he addresses the earthly throne as well as the cosmic. Reciprocal coition is the mark of the natural political order of the universe, of mutual alliance rather than tyrannical violence.

Magnets are amiable Elizabethan subjects, enjoying the peaceful love of their sovereign, not inordinate Spanish lust. Magnetic coition is not passionate and turns away from the violent transports of sexuality. Describing the parts of the Earth or of a terrella, Gilbert says that they "are in accord and enjoy neighborhood with each other: there is in them all mutual love, undying good-will." Having led us into the "inward depths" of the Earth, he wishes us to behold a coition that is

> not indeterminate and confused, it is not a violent inclination of body to body, not a mad chance confluence. Here no violence is offered to bodies, there are no strifes or discords; but here we have, as the condition of the world holding together, a concerted action,—to wit, an

> accordance of the perfect, homogenous parts of the world's globes with the whole, a mutual agreement of the chief forces therein for soundness, continuity, position, direction, and unity.

Gilbert's universal vision of "mutual love, undying good-will" ultimately informs his rejection of violence. Yet iron, the hardest of metals, needs a "hard master," one who must reconcile hardness with love.

2

Wrestling with Proteus

. . . the nature of things betrays itself more readily under the vexations of art than in its natural freedom.
—Bacon, *The Great Instauration*

Francis Bacon gave as much thought to political mastery as to the mastery of nature. The son of Queen Elizabeth's Lord Keeper, Bacon gradually rose through the practice of law to the office of Lord Chancellor and was, at the height of his career, second only to the king in official power. His immense ambitions extended far beyond his own advancement; he used his eloquence to try (unsuccessfully) to persuade his royal patron, King James I, to sponsor a vast state-supported project to investigate nature on novel lines. Bacon wanted a new kind of natural philosopher who would not merely gaze respectfully at nature, but would engage her in an intense mutual trial. Although there was no love lost between him and Gilbert, both appreciated that struggling with nature yields powerful insight. Like a careful physician, Gilbert worried whether acids might "poison" his lodestones. Bacon expressed no such fears as he dissolved his magnets in acid, then ground them to powder to test whether their attraction would survive. He burned his magnets whole, or raised them to the dizzying heights atop St. Paul's Cathedral.

Bacon's vocation as a lawyer and judge bears on his investigations. As queen's counsel, later solicitor general and attorney general, Bacon was versed in all aspects of the examination of witnesses and the evaluation of testimony. In his private speculations, he marshaled a similar investigatory apparatus that would penetrate into Nature, whose "genuine forms . . . lie deep and are hard to find." Nature is put on trial and examined through her own testimony. This regular procedure of questioning is called "examination upon interrogatories" in English law, and Bacon claims, "(according to the practice in civil causes) in this great plea or suit granted by the divine favor and providence (whereby the human race seeks to recover its right over nature), to examine nature herself and the arts upon interrogatories."

Bacon used the term "vexation" to describe this interrogation; in his usage the word means troubling, afflicting, or harassing. In many passages, the tempered, inward sense of vexation distinguishes it from the sheer brutality of torture. Later, critics who accused Bacon of putting nature on the rack ignored this distinction. Although Continental law (following Greek and Roman precedents) used judicial torture to verify circumstantial evidence, torture was not utilized in English common law, which relied on the jury to establish legal proof. Bacon was well aware of the lack of legitimacy of torture; though he was present when torture was used in extraordinary proceedings to uncover plots against the king, he was reluctant and expressed reservations. In his legal treatises, Bacon made no mention of any royal prerogative to torture, although he was in other respects "the king's man." He only notes that "in the highest cases of treason, torture is used for discovery, and not for evidence," in order "to identify and forestall plots and plotters." During this period, Edward Coke, Bacon's great legal rival, had no qualms about using torture, though after Bacon's death Coke stated that it was contrary to English law. Both men acted under the command of James, who, as king of Scotland and before his accession to the English throne, had presided over witchcraft trials. A contemporary account relates that James "took great delight to be present" at the examination of accused witches, including their "most strange" tortures. His ministers had to follow his royal commands.

Even if it is not torture, vexation indicates strong, probing interrogation. Bacon justified such intense trials by the urgent need of suffering humanity for relief. His religion hinged upon the exaltation of charity, especially the healing of illness and infirmity. In Bacon's view, Jesus as savior was ever mindful that "the body of man stands in need of nourishment, of defense from outward accidents, of medicine. . . . All his miracles were for the benefit of the human body, his doctrine for the benefit of the human soul. . . . There was no miracle of judgment, but all of mercy, and all upon the human body." To imitate this divine charity, man must aspire "to a goodness not retired or particular to himself, but a fructifying and begetting goodness." This requires an arduous quest into the depths of Nature, even into "the deeps of Satan, that [the scientist] may speak with authority and true insinuation. Hence is the precept: Try all things, and hold that which is good: which induceth a discerning election out of an examination whence nothing at all is excluded."

When he comes to clarify these matters, Bacon goes beyond legal metaphors. He does not turn so much to records of current experiments, perhaps because he felt they were still only in an initial and inconclusive stage. Instead, he turns back to the oldest myths to find images and stories that might be adequate. In part he does this to speak to an audience steeped in classical literature, for whom the emergent science was alien and perplexing. But even more, Bacon suspected that these ancient fables contained, under a veil of allegory, crucial clues to the future.

Bacon illuminates the struggle for the secrets of Nature in his treatment of the story of Proteus, the Old Man of the Sea. Homer's *Odyssey* tells of the Greek hero Menelaos, who, while trying to return home from the Trojan War, was mysteriously becalmed off the coast of Egypt. Sensing that he had displeased the gods, Menelaos sought help from the ocean-dwelling Proteus. Proteus was a shape-shifting immortal; in Homer's version, his daughter pitied Menelaos and advised him to creep up on her sleeping father and hold him fast, for he would try to escape by transforming himself into every animal and element. The hero can only relax his grip when Proteus resumes his normal shape; then Proteus will respond to his questions.

In his version, Bacon notes that Proteus, as the herdsman to Neptune, was a prophet, "the messenger and interpreter of all antiquity and all secrets." Bacon interprets Proteus as "Matter—the most ancient of things, next to God," and Bacon's fable relates how experiment reveals the secrets of nature. Bacon explains that "the vexations of art are certainly as the bonds and handcuffs of Proteus, which betray the ultimate struggles and efforts of matter." This interrogation requires handcuffs and chains, but it is not a scene of torture. The hero's struggle with Proteus is *mutual,* testing both of them. The seeker is not self-sufficient, not self-fashioning; everything depends on an encounter with forces beyond the self. In Homer, Menelaos and his companions wrestle bare-handed with the Old Man, whereas Bacon's "skillful Servant of Nature" employs "mechanical" aids: handcuffs and chains. What had been in Homer a trial by combat, a test of strength and courage pitting men against a god, becomes in Bacon less dependent on personal valor. Bacon speaks of "*any* skillful Servant of Nature," emphasizing that a unique hero such as Menelaos is no longer called for; many seekers can approach and constrain the god to answer (figure 2.1).

Nevertheless, the Servant of Nature must be qualified and must endure. In Homer, Proteus's daughter had prepared Menelaos for his trial. There is a time and due measure to hold Proteus passing through his manifold shapes. Though Bacon does not include the daughter, in his version, Proteus tries the seeker as much as they try him; their ordeal is mutual and has an appointed ending, after which force must cease. This occurs when Proteus has resumed his original form; the seeker's vexation is of no avail if he cannot recognize that primal form and release Proteus into his original freedom. When Proteus resumes the uses of words and questions his questioner, the hero must relax his grip. Holding tight is not enough; at last the god must be left free to respond and ask a counterquestion of the seeker.

Bacon's Servant of Nature is grappling with a lower deity, with Matter, and not with God directly. Unlike some of his contemporaries, Bacon views Proteus as godlike, not an image of fallen nature. Only God can annihilate. The Servant of Nature acts as if he could reduce matter to nothing, but knows that his strength cannot really destroy. Still, his force induces manifold changes, which finally circle back to their beginning and disclose the secret. As Bacon says, "all that man can do is to put together and put asunder natural bodies. The rest is done by nature working within."

Bacon emphasizes the time set for the confrontation with Proteus. It is high noon, which Bacon identifies as the exact moment of creation. The Servant confronts matter in all

Figure 2.1
Giulio Buonasone, Proteus and Aristaeus (Achille Bocchi, *Symbolicae quaestiones,* 1574).

possible forms—the form of forms—not one solitary, isolated form. The moment of the struggle with Proteus is always the moment of creation, charged with divine energies ready to disclose their secret to the properly prepared seeker, who does not merely behold them as a spectacle, but actively unleashes their tumult. The shifting forms of Proteus are without significance for the unprepared beholder, but are harbingers of true knowledge for the rightful Servant of Nature. This is a test of strength and discernment on the highest level, only fitting for those rightly prepared, as both Menelaos and the Servant have been. Matter is the eldest child of creation and answers the ordeal of experiment without prevarication. The majesty of Nature requires commensurate treatment. In unpublished writings, Bacon treats Nature as a captured queen, but rather than treating her as a slave, he calls for his "dear, dear son" to unite in a "chaste, holy, and legal wedlock . . . with things themselves." Nature is raised to the status of a wife.

Though Bacon felt that God had enjoined man to subject nature to penetrating interrogation, he is aware of cases in which experiment is cruel or inhuman.

> For to prosecute such inquiries concerning perfect animals by cutting out the fetus from the womb would be too inhuman, except when opportunities are afforded by abortions, the chase, and the like. There should therefore be set a sort of night watch over nature, as showing herself better by night than by day. For these may be regarded as night studies by reason of the smallness of our candle and its continual burning.

Bacon sees the danger that intemperate experimentation might elicit misleading or distorted responses from nature; torture is futile because it tends to elicit false or garbled confessions. As Gilbert had noted, excessive fire tends to distort and falsify the results. The eliciting of nature's secrets requires a proper respect for her nobility and commanding sway. Bacon reminds us that "you may deceive nature sooner than force her." Since she cannot be fooled, "nature to be conquered must be obeyed." Man cannot enter nature's inner courts without confronting her inherent greatness, but must be ready for an intense and confusing struggle. Then everything depends on the soul of the seeker.

II

Desire and Science

3

The Wounded Seeker

Nor is that other point to be passed over, that the Sphinx was subdued by a lame man with club feet; for men generally proceed too fast and in too great a hurry to the solution of the Sphinx's riddles; whence it follows that the Sphinx has the better of them . . .

—Bacon, "Sphinx, or Science"

Seekers must refine themselves before they can begin to question nature rightly. This is not merely ritual or symbolic purification. They must be tempered in the exact sense that a blade is tempered by being heated red-hot and then plunged into cold water: the extremes of heat and cold purge passions that would weaken the blade. Their sensibilities must be scourged of common human "knowledge drenched in flesh and blood"; they must achieve a kind of chastity immune to the seductions of mere appearances. Bacon warns "every student of nature . . . that whatever his mind seizes and dwells upon with peculiar satisfaction is to be held in suspicion." Otherwise, the student will mistake his own illusions for God's secrets. The secretive creator tests our discernment to the utmost, for as Solomon said, "The glory of God is to conceal a thing, but the glory of the king is to find it out."

Nothing can be excluded from this search, neither serpents nor poisons nor infection. God will protect the seekers from any pollution and even commands them to search fearlessly. "Neither let any man think that herein he tempteth God; for his diligence and generality of examination is commanded; and *God is sufficient to preserve you immaculate and pure.*" Bacon takes the Scripture to reassure the seekers of their divine license. Yet how are they to avoid blind presumption? The human mind is not a faithful mirror of the world, as Aristotle taught, but "an enchanted glass, full of superstition and imposture, if it be not delivered and reduced." Our vision is distorted and corrupt, ruled by idols of common delusion. Science can only advance if there are means to deliver the mind from its illusions.

The story of the Sphinx gives the key. The winged virgin lay in ambush near Thebes, ready to kill with her claws any traveler who could not answer her riddle: What is born four-footed, afterwards becomes two-footed, then three-footed, and, finally, four-footed again? The Sphinx is a symbol of science, who "being the wonder of the ignorant and unskillful, may be not absurdly called a monster." Here "monster" means a marvel or prodigy, in the usage of Bacon's time. Aristotelian science had neglected monsters as exceptions, deviants from the normal course of nature; critical of this tradition, Bacon identifies the new science as itself a wonder that does not exclude deformity.

Each feature of the Sphinx is an image of an aspect of science. "In figure and aspect it is represented as many-shaped, in allusion to the immense variety of matter with which it deals. It is said to have the face and voice of a woman, in respect of its beauty and facility of utterance. Wings are added because the sciences and the discoveries of science spread and fly abroad in an instant. . . ." Despite certain human features, science is multiform and more than human. The womanly beauty of its voice and face connects it to the Sirens, whose alluring song Bacon will treat as a figure of Pleasure. Rather than tormenting with overwhelming pleasure, the Sphinx seizes its victims with its hooked claws, signifying scientific questions that seize the mind with agonizing fascination. Practical science is not distant and impersonal awareness, but vivid, even painful, realization.

Although he points out the pains, Bacon is also aware of the pleasures that accompany scientific fascination. The "hard questions and riddles" came originally from the Muses, and were not cruel until the Sphinx got them. At first, science is pure pleasure, the delight of unfettered inquiry. However, when contemplation inevitably leads to practical application, the cruel dilemmas appear, especially when reasons of state require immediate action. Bacon always remembers the political dimension of science, for it gives new powers to humanity. He notes that Augustus Caesar used a Sphinx for his seal, since Augustus, too, had solved great riddles in political life. After the murder of Julius Caesar, Augustus was a second Oedipus, who saved Rome from the sphinxlike peril of civil war; he grasped the new secret of political power and forged the Empire. The scientist who solves the Sphinx's riddle is "born for empire," as Bacon nakedly puts it, an empire not only over nature, but also over man.

In Bacon's account, the scientist is not an unpolitical servant, meekly delivering his discoveries to his masters, but takes responsibility himself. Those who wield such power should be prepared and purified. The riddles of the Sphinx pose a mortal threat, as well as an immense prize: "distraction and laceration of mind, if you fail to solve them; if you succeed, a kingdom." Indeed, she tears to pieces those who fail.

Oedipus can enter into the full laceration of the riddle because of his own lameness. Rather than crippling disabilities, these wounds are the very means through which he finds the solution to the riddle: his lameness slows him down and lets him grasp what others hurry past. His Greek name suggests "the one wounded in his feet" and, also, "he who sees and knows as he stands on his two feet." As Oedipus faced the Sphinx, her riddle must have struck a deep resonance, for it concerned the feet of a man and their number. Oedipus, more than any other man, had to count on his feet, beginning with the single pierced-together limb his father inflicted on him, then the four on which most of us begin to crawl, and the two on which we walk. Oedipus answers "readily," Bacon says. The abused and exposed child became the man who solved the famous riddle. Bacon is silent about the darker aspects of the story: parental mutilation and exposure, parricide, incest, and self-blinding. I think that Bacon hoped that the new scientists, Oedipus's spiritual heirs, would in their greater wisdom avoid the tragic crisis that overwhelmed their ancestor. In Bacon's account, Oedipus is triumphant; his wound allows him to approach the Sphinx so slowly and warily that he can overcome her.

Bacon's connection of lameness with insight applies to other scientists; both Newton and Einstein achieved their greatest work because they could think excruciatingly *slowly*. Indeed, scientists have periods of illness or physical disability in childhood far beyond the proportion in the general population. Bacon may also have intended a reflection on the extraordinary destinies of certain lame people, including King James and Bacon's own brother, Anthony. Francis considered Anthony to have a more active and able mind than his own, despite his brother's "impotent feet." The new science mirrors Anthony's combination of lameness and insight. Francis also judged himself weak in health, but understood, with his brother and Oedipus, the paradoxical advantage of deformity, which spurs a desire to escape from scorn. Deformity also deflects envy; impotence inspires trust. More darkly, those whom nature has mangled may seek their revenge on her; they may be void of natural affection. The character of Shakespeare's Richard III is as twisted as his body is misshapen.

Only a few deformed persons avoid being inwardly twisted. They are "extreme bold" and sometimes prove excellent persons, among whom Bacon includes Socrates. The deformed are not healed through faith; in exceptional cases, their deformity helps them as they contend with it. Bacon's Oedipus does not blame his father, but could blame nature. However, Oedipus overcame the inclinations of his pierced body by "discipline and virtue." He "readily" recognizes himself in the Sphinx's riddle, *because he does not fit it.* Having begun with only one "leg," Oedipus can grasp a truth hidden to all others by overfamiliarity.

Oedipus's wound is a path to a new kind of knowledge. It turns disability into a source of profound insight. Bacon specifies that Oedipus killed the Sphinx; in the ancient sources she kills herself. The impulse to have revenge on nature has turned to the eradication of human suffering. The Sphinx threatens all men with her riddle, unless it is solved. The specter of death stands at the gate of every city. By killing the Sphinx, not Nature, Oedipus transforms vengeance into an ennobling attack on the monstrous riddles that afflict men.

To go beyond the case of the individual seeker, Bacon retells the story of Prometheus, the titan who stole fire and gave it to mankind, and was punished by an eagle that gnawed his liver. By the "school of Prometheus" Bacon means "the wise and fore-thoughtful class of men," those who by relentless investigation seek to better the human condition. But while they work so arduously, they torment themselves with gnawing cares. Their inward suffering mirrors their vexation of Nature. Only Hercules, signifying fortitude and constancy of mind, can save them from their torments, as the ancient myth prophesied. Yet Prometheus was a criminal; he tried to rape the goddess Minerva, to possess divine wisdom through mere force. This was part of his bold plan to bring divine secrets to men, which cannot avoid exciting inordinate desire. The restless, erotic quality of Promethean striving must be bridled and chastened. Prometheus also overstepped the boundary between the realms of divinity and science; a certain chastity is required if science is not to be contaminated with fable, or if religion is not to become heretical. Otherwise, science may spend its strength on alluring but empty visions that lack power to help humanity.

This restless striving finds its proper object in unraveling nature's secrets, but the ardor of the seeker needs tempering. As later chapters will explore, those secrets are encoded in their own symbolic language, not exposed in common human tongues. Its solution requires not just a few, rare minds, but, rather, a succession of many workers all arranged in what Bacon calls a "machine"—a highly articulated organization that requires many persons of varying capacities to penetrate the code. This vast collective undertaking Bacon calls the "games of Prometheus," recalling the legendary torch races that honored the fire-bringing titan. Solitary, limping heroes like Oedipus prepare the way for the waves of runners that follow them. In this relay race, the individual participants pool their desires toward these new public goals and away from the divisiveness of private eros.

Only those whose sensibilities have been disciplined are capable of breaking the code; the process of decryption involves so much humiliation and painful trial that it chastens and prepares the decoders. Because of this safeguard, which Bacon ascribes to the secretiveness of God and nature, any attempt to use brute force to penetrate the secret not only will fail, but also may recoil disastrously, for "force maketh Nature more violent in the return." If so, Bacon has restored hope to humanity, not the blind hope that the ancient Prometheus gave men. Scientific hope relies on "a true and legitimate marriage between the empirical and the rational faculty." Where Prometheus tried rape, science will enjoy marriage.

4

The Creatures of Prometheus

Here is my world, my universe.
Here is where I feel myself to be.
Here are all my desires
In bodily form,
My spirit divided a thousandfold
And whole in my beloved children.
—Goethe, *Prometheus* fragments

In retelling the stories of Oedipus and Prometheus, Bacon indicates the depth of the purgation needed to approach nature's secrets without haste or delusion. Yet these arduous ordeals must not weaken the seeker's desires, but turn them in new directions. Bacon uses a sequence of myths to describe these transformations of eros. They begin with Vulcan, who, like Oedipus, was wounded by his parent and limps. As the god of the forge, he is the original possessor of the fire that Prometheus stole. Vulcan's lameness is the price of his mastery as artificer. His erotic life is also lamed; though he marries the goddess of love, she cuckolds him.

Vulcan was not born from sexual intercourse, and, compared with the other gods, had few children. Instead, he crafts automata and self-moving tripods, amazing mechanisms whose abilities mimic those of intelligent beings. He even made a human being, Pandora, who was sensuality incarnate. As with Prometheus, Vulcan's desire was aroused by the goddess of wisdom, whom Vulcan tried to force when she refused his advances. He was not successful even in rape; "in the struggle which followed his seed was scattered on the ground; from which was born Ericthonius, a man well made and handsome in the upper parts of the body, but with thighs and legs like an eel, thin and deformed: . . . he, from consciousness of this deformity, first invented chariots, whereby he might show off the fine part of his body and hide the mean."

Bacon interprets this assault as an image of scientific artifice attempting "by much vexing of bodies to force Nature to its will and conquer and subdue her." Though his rape fails, Vulcan's seed still has generative power. It leads to the chariot and other technological advances that are estimable, yet lame compared to true scientific insight. The followers of Vulcan, votaries of fire and force, lust after practical results; they ignore the rightful wooing of Nature, treating her as a slave to slake their desires. Similarly, Atalanta is distracted from her race by the desire to pick up golden apples strewn in her path; scientists are tempted to turn toward the immediate gratification of profit. Bacon recognizes the difficulty of channeling the scientists' desires toward a higher but less immediate goal, which he identifies as *light*.

Accordingly, his treatment of myth critiques common eros and redirects it toward this new goal. This critique reaches past conventional sexual roles, as important as they may be. Phrases like "the sons of science" show that Bacon shared the masculine prejudices of his time, as did the brilliant Queen Elizabeth. Such limiting beliefs were already old and deeply rooted; their effects pervaded society, then as now. Bacon criticized marriage because the sweetness of family life distracts men from devoting themselves to serving the commonwealth. His own marriage was childless and probably unhappy, though speculation about his sexual identity lacks definitive evidence. In any case, Bacon probed a Protean realm of feelings that cannot be confined to conventional masculinity and femininity. He understood that if science is to sound the depths of nature it must also delve past appearances to wrestle with hidden desires. This is especially true if science sets out to reshape eros itself. After all, Bacon's Cupid represents the impersonal attraction of atoms, not a roguish archer. Ironically, his Cupid has no parents but is "an egg laid by Night"; the erotic impulse comes from an atomic abyss.

Bacon's darkest suspicions about passion emerge as he tells the story of Dionysus. Eros is a consuming fire; Dionysus's mother was scorched when she wanted to behold Jupiter's naked form. After that, the king of gods limped as he carries the unborn child sewn in his thigh, sharing with Oedipus and Vulcan this creative deformity. Jupiter's strange maternity reflects the sexual ambiguity of his child, "for every passion of the more vehement kind is as it were of doubtful sex, for it has at once the force of the man and the weakness of the woman." The labyrinth leads to a Minotaur born of mingled animal and human lusts. Bacon treats the career of Dionysus as a crescendo of unremitting evil, for desire is insatiable and pitiless toward everything that stands in its way. Passion disregards honor and never dies, for when it seems extinguished, it suddenly flares up again. In the end, insane passion leads to superstitious rejection of true religion and science.

This climactic accusation seems to be a final revelation of the inimical relation between passion and science, except that a quiet anticlimax undercuts it. "Lurking passion or hidden lust" might lead to outstanding deeds practically indistinguishable from those that come "from virtue and right reason and magnanimity." If so, the wellsprings of desire yield forces whose power Bacon cannot neglect or discount. He does not merely give a puritanical rebuke to pleasure, for he is aware of radical problems that afflict the school of Prometheus. The eagle of anxiety gnaws their livers, leaving them perturbed and afraid. Then what advantage have the Promethean scientists if they are constantly ravaged by worry? Bacon's reply is guarded; he refers to the coming of Hercules, "that is, fortitude and constancy of mind," to rescue Prometheus. This fortitude is a condition, "which being prepared for all events and equal to any fortune, foresees without fear, enjoys without fastidiousness, and bears without impatience." In rehabilitating pleasure Bacon emphasizes that the acquisition of scientific knowledge should make us happy and fearless, not careworn and anxious. Without an enlarged capacity for pleasure the scientists' sensibilities might be too constricted to achieve true largeness of spirit. His example is Solomon, the wisest of kings, whose collection of natural history Bacon holds up as a model for King James. In order to motivate his suspicious monarch to patronize science, Bacon encourages such largeness of spirit. Bacon emphasizes Solomon's keen desire to pursue his prescient scientific activities. Solomon also enjoyed physically his many loves, who (in contrast to the Biblical account) never led him to worship false gods.

The scientists must remain open to the heights of pleasure not only to ensure that their sensibilities remain unconstricted, but also for a strictly scientific reason: they must not avoid exploring the phenomena of pleasure out of fastidiousness or a fear of pollution. In the alluring or the disgusting, as well as in the indifferent, may lie crucial discoveries. Bacon's version of "The Sirens; or Pleasure" emphasizes the possible dangers of this project. So powerful is the Sirens' song that, like the white bones of their victims shining from afar, even "the examples of other men's calamities, however clear and conspicuous, have little effect in deterring men from the corruptions of pleasure." Ordinary men, like Ulysses's crew, had better stop their ears if they cannot master such temptations.

Even minds of loftier order must venture cautiously into the midst of pleasure. They should behold pleasures as observers, like Ulysses tied to the mast of his ship. But where Ulysses used philosophy, Orpheus used religion. This remedy is best since Orpheus "by singing and sounding forth the praises of the gods confounded the voices of the Sirens and put them aside: for meditations upon things divine excel the pleasure of the sense, not in power only, but also in sweetness." Bacon's warning against mixing religious and scientific matters does not exclude scientists' imitating Orpheus by raising their discourse to the exaltation of divine mysteries. The highest access of scientific inquiry is not the anxious confrontation with the Sphinx of practical choice; it is the ravishing sweetness of knowing the secret causes behind Nature. Orpheus rejoined the Muses not by avoiding pleasure or experiencing it guardedly, but by raising his sweet song to the gods. Similarly, the scientists experience unpolluted pleasure of the highest kind when they enter the realm of the Muses. Orpheus reclaimed pleasure for a musical science that can enchant and exalt the spirits of men, not by denying their passions, but by raising them to a higher key. Bacon seems to envisage something like the rapture that Einstein experienced as he struggled with "God's thoughts": an intense, almost physical thrill more fascinating than the pleasures of ordinary life.

Yet even Orpheus finally could not escape fatality. After the death of his beloved wife, Euridice, Orpheus had virtually succeeded in bringing her back to the earth when his impatience overcame him and he turned around to gaze at her, violating the command of the gods and forcing her to return to the underworld. Bacon interprets this as "curious and premature meddling," as if an impatient scientist were to disturb a crucial experiment when it was on the brink of succeeding. Here a fateful experiment is under way, for restoring the dead body to life is the climax of "the restitution and renovation of things corruptible." After Euridice's second death, Orpheus emerged as a benign political figure, teaching "the peoples to assemble and unite and take upon them the yoke of laws and submit to authority, and forget their ungoverned appetites, in listening and conforming to precepts and discipline." He was a teacher of rational political science, whose eros is suffused with "the love of virtue and equity and peace." This rational Orpheus turned toward the more subtle defeat of mortality through a transformation of political life.

In the ancient story, Orpheus rejected woman and family and introduced a new kind of eros: the love of boys. In contrast, Bacon's Orpheus enchants the world to a new height of peace. Wild beasts, "putting off their several natures," forget their quarrels and ferocity, "no longer driven by the stings and furies of lust, no longer caring to satisfy their hunger or to hunt their prey." Orpheus is able to touch these feelings of love and virtuous action without stirring discordant desires because of his sorrowful mood, which, Bacon notes, marks philosophy after it failed in its first attempt to conquer death. As wild beasts and men give over their savage desires for conflict, a purified realm of feeling, dominated by scientific concern, banishes war and strife. Nevertheless, Dionysus incites the women of Thrace to counter this dangerous attack on himself as Desire personified. Orpheus cannot withstand the "hoarse and hideous blast" that the women blow on their horn. The peaceful fellowship Orpheus enjoyed with his animals is broken; the women tear the singer to pieces and scatter his limbs. Their rejected desires overwhelm Bacon's dream of a new eros. Although Orpheus failed again, we are invited to take up his lyre. As with Prometheus, Bacon's vision shifts from individuals to whole nations of seekers who will carry the torch.

Bacon embodies his hope in the Muses' sacred river Helicon, who "in grief and indignation buried his waters under the earth, to reappear elsewhere." In Bacon's cyclical vision, kingdoms flourish, then sink to war and barbarism, leaving letters and philosophy "so torn in pieces that no traces of them can be found but a few fragments, scattered here and there like planks from a shipwreck." But this dissolution is followed in the cycle by the waters issuing forth again elsewhere, in other nations. Bacon looks to a new hope, a scientific utopia that will harmonize desire and science. There, the rising tides may carry Orpheus's hopes to completion.

5

The New Eros and the New Atlantis

For there is nothing amongst mortal men more fair and admirable, than the chaste minds of this people.
—Bacon, *New Atlantis*

Besides these discourses drawing on ancient myth, Bacon also shaped a modern myth in the *New Atlantis*. In it he describes the profound alterations in human desire that he anticipates will accompany the growth of science. Bacon situated his vision of an advanced scientific civilization on Atlantis, the fabulous island of Plato's mythical empire, which was rich in technological wonders, but finally became arrogant and aggressive before its downfall. Bacon's New Atlantis must avoid these fatal flaws.

Bacon hides his island from European eyes and gives it a Hebrew (rather than Greek) name that conjures up Biblical associations: Bensalem, the son or abode of peace. Its inhabitants rescue the narrator and his shipmates after they are wrecked. The travelers are amazed at the charity and generosity of their rescuers and even wish to remain there, rather than return home. It seems to them a paradise, far in advance of Europe both in technology and humane values. Although much of the *New Atlantis* concerns the wonders of science in Bensalem, there are remarkably extended passages on the erotic life of these scientific islanders.

Every town has "Adam and Eve's pools," in which a friend of the prospective bride and another friend of the prospective groom are permitted to see the bride and groom separately bathe naked before a marriage is contracted. The custom is justified "because of many hidden defects in men and women's bodies." It is explained by Joabin, a Jew said to be "a wise man, and learned, and of great policy, and excellently seen in the laws and customs of that nation." He says that the prospective pair's friends might be able to discern hidden defects in their respective bodies; it is not clear whether these "defects" refer to what the friends consider erotically repulsive or perhaps signs of infertility.

Many utopias regulate the breeding of human beings on the model of animal husbandry. From the traditional point of view, such practices are outrageous. Indeed, the Bensalemites have a grave man or matron present the naked man or woman to the prospective partner, thus investing the moment with a solemnity that will ward off laughter or inappropriate outbursts of desire. These practices are accompanied by a mandatory waiting period of one month between first meeting and marriage, as well as by sanctions against marriage without parental consent.

When Bacon advises every student of nature that "whatever his mind seizes and dwells upon with peculiar satisfaction is to be held in suspicion" he means to chasten overfondness for some pet idea; the habitual practice of self-doubt is a crucial guarantee of the honesty of the scientist. In contrast, the alchemists "grow old and die in the embraces of their illusion," as if their grand designs were deceitful jades. Yet eros does not necessarily wane if it is suppressed, but may well become more furtive and stronger. At the pools, the eyes of friends guard erotic choice not through puritanical repression but judicious exposure. This is only possible if those friends, and indeed the whole of society, are incorruptible. The immense difficulty, not to say incredibility, of such purity arising and sustaining itself is a central problem of this utopia. Indeed, the Bensalemites hide from the rest of the world, on which they spy, but from which they remain invisible. As great as the virtue of the Bensalemites may be, they do not underestimate what might happen if they opened their doors to the world beyond.

Atlantean science includes erotic matters in its universal scope. The crucial interview with Joabin is interrupted by the rare visit of a Father of Salomon's House, a member of the highest scientific society in Bensalem. Bacon's timing suggests that this conversation about erotic matters is interrupted by the arrival of the archscientist, whose revelations will make clear the context of these extraordinary customs. That Father explains that they know how to make copulations of different kinds of animals. These novel hybrids are not haphazard; the erotic lives of these species are rigorously controlled. Although he is speaking about the production of hybrid animals, nothing he says prevents such techniques being applied to humans. This preoccupation with generating new species runs throughout Bacon's project. He foresees "Deceptions of the senses. Greater pleasures of the senses. . . . The altering of complexions, and fatness and leanness." His scientists are interested not only in optimization of breeding, but also in altering the breeding stock itself, down to its sensual awareness. They could alter the body and alter erotic pleasure; they could produce "instruments of lust, and also instruments of death," both of which could shake the world.

Eros is one of the deepest held, most secret parts of human nature; touching the wellsprings of desire may make people happier, but it may also set them adrift from their traditional customs. Despite these possibilities, Bensalemite social order is extremely conservative. They have opted for an angelic chastity that rejects the dissoluteness of European mores. The reasons for this choice are not entirely clear. One is led to expect that the Bensalemites are following the precepts of revealed religion, which lie beyond the test of science. Bensalemite science was highly developed even before their special revelation of Christian doctrine; there, science was able to approve and verify divine miracles, avoiding any conflict with religion. Indeed, their science is so powerful it could have manufactured the miracles wholesale. The scene of presumed revelation might have been a noble lie staged by pious but devious scientists, except that they expressly forbid such deceptions. As it is, devotion to science is tinged with religious awe. The Father of Salomon's House has the presence of a great prelate, down to his benedictions to the crowds who seek his blessing; the priest seems a humble assistant of the scientific hierarchy. The king has quietly ceded to Salomon's House control over the vital secrets their science discloses. All this implies a close relation between science and Christian humanism; Bacon expected that classical and Christian learning would continue to flourish, and he depended on them as the foundations of humane sensibility. He would have been amazed to see how both have languished.

The new eros expresses the needs of science. Although the scientist cannot keep a "mistress" among theories, he needs to maintain a fervent love of science to sustain him through many trials and disappointments. In the New Atlantis, mental chastity accompanies sexual moderation. Since the scientist acts as a member of a vast machine, scientific eros needs to be attached to institutions, to the whole company of Promethean racers. It is scarcely surprising that such a demanding commitment might, in most humans, leave not much eros left over for private, conjugal life. Here one is reminded of Bacon's Orpheus, who turns from the perishable love of women to his new eros and his pious, ecstatic science.

In *Gulliver's Travels,* Swift made fun of the scientists on the flying island of Laputa, so lost in their abstruse studies that they employ flappers who prod them to eat. The contemptuous wives of these Laputan projectors seek from gallants the sexual solace that their scientific husbands are too distracted to give. The name of their preposterous flying island means whore in racy Spanish. Their scientific ecstasy is deflated by the neglected claims of ordinary eros. Though Swift mocks the abstractedness of the Laputan scientists, he testifies to the fascination such projects can have, and the surprising way they can rival the ordinary charms of love. In Bensalem, however, science is altogether triumphant and the narrator is brought to confess abjectly that, not only in material power and prosperity, but in all respects, Bensalem is more righteous than Europe. There is no dissenting voice heard or even suggested, and the women, too, are represented as part of the uniformly contented chorus of citizens. However, certain aspects of family life give an insight into the way familial and personal eros is affected by the prevalent passion of the Bensalemites for scientific light and fruit.

The narrator is much affected by the "Feast of the Family." In this custom, the father of a family with thirty living descendants receives elaborate and stylized homage from his family and even from the king, who acknowledges the father's remarkable contribution to the population of the realm. This father sits on a raised dais like a monarch and eats alone, joined only by a son "if he hap to be of Salomon's House." The custom ritualizes the ceremonial scenes of blessing drawn from the Bible, allotting such magnificence not to merit or rank or service, but only to fecundity. During these ceremonies, the mother of the family is installed out of sight "in a loft above on the right hand of the chair, with a privy door, and a carved window of glass, leaded with gold and blue," but only if she is the mother of all thirty. The female birth-giver is hidden, but her place of honor is above the masculine progenitor. Despite patriarchal bias, Bacon glorifies her secret vantage point.

The various kinds of eros each have their own special children. Alchemists die in fruitless embraces, but the true scientist is rich in living offspring. In the Feast of the Family, the Father is not only the scientist, for this ceremony is specified for all numerous families, reserving some extra dignity for those of rank, and particularly those of Salomon's House. The larger implication of the feast is the need to encourage such large families. Clearly there is some danger that population might dwindle and Atlantis lack sufficient inhabitants, an important issue if it is to remain economically self-sufficient. This seems to be on the narrator's mind when he asks "whether they kept marriage well; and whether they were tied to one wife? For that where population is so much affected, and as with them it seemed to be, there is commonly permission of plurality of wives." In the usage of Bacon's time, "affected" implies "desired, aimed at." Thus the narrator had expected that polygamy *would* be permitted since he noticed that the island was so underpopulated.

The reasons for this underpopulation are unspecified and perplexing. The habitable area is not excessive, since the island is no bigger than England and France put together; the soil is fertile and food is plentiful. The climate is mild; scarlet oranges grow there. The mysterious pills that heal the sailors indicate medicine far more advanced than in Europe. Nor does one expect infant mortality to trouble a society that prepares "Water of Paradise" for health and longevity. The Bensalemites have even acquired control over fertility and sensual pleasure. There has been no war or civil strife for nineteen hundred years, giving ample time to populate the island. In all that time, only thirteen foreigners returned from Bensalem, which is scarcely troubled by outside incursions.

The only remaining explanation for the underpopulation is some alteration of eros so deep that it cannot be remedied by science, neither by augmenting fertility nor by heightening sexual pleasure. Evidently families are less numerous, even though scientific fruitfulness is exemplary. The inference is that so much eros goes into the Atlantean's scientific projects that sexual desire wanes and families shrink. Joabin criticizes the Europeans for moral laxity, avoiding marriage or marrying late, "when the prime and strength of their years is past." He implies that these are not the problems of Bensalem, where the minds of the people have remained chaste. Both the Bensalemites and the Europeans face the same problem, for different reasons: a diminution of the birth rate. In the case of Europe, this is the result of excessive eros channeled to debauchery, while in Bensalem, eros is absorbed by the chaste raptures of science. At least the institution of Adam's and Eve's pools would be inconceivable in Europe, which Joabin depicts as a furnace of unlawful lust, which "if you give it any vent, it will rage."

This is also deeply connected with the peacefulness of Bensalem, the "son of peace." The islanders abhor war and have found ways of living together in concord. It seems, then, that this peaceful disposition has been cultivated as their erotic life has been tamed. And now, the Fathers of Salomon's House are permitting these great secrets, so long hidden, to be carried into the outer world. The Atlanteans seem confident that Europe will embrace the scientific miracles of Bensalem, which will conquer the world peacefully. Clearly they run a great risk, for even with their technical prowess it is not certain that they could withstand all-out attack from a greedy and corrupt Europe. At this point, Bacon's fiction breaks off and it is unclear whether he regarded the work as finished or even finishable. The question remains whether human nature will be able to be tamed so that the powerful secrets of Atlantis will prove to be a blessing or a curse.

Although the Father of Salomon's House disclosed unheard-of wonders, he did not explain how they were accomplished. These alluring possibilities inflame desires that might otherwise be overchastened. Icarus perished from flying too high, but insufficient desire poses a greater risk "because in excess there is something of magnanimity—something, like the flight of a bird, that holds kindred with heaven; whereas defect creeps on the ground like a reptile." Yet those untouched by scientific light can still grasp its fruits. Bacon notes that "when the Sphinx was subdued, her body was laid on the back of an ass: for there is nothing so subtle and abstruse, but when it is once thoroughly understood and published to the world, even a dull wit can carry it." Once revealed, scientific secrets become banal, their powers open to corrupt rulers. Bacon surely knew from his own failures with King James how difficult it would be to guard scientific powers from political abuse or neglect. After all, the king had mocked Bacon's scientific project as something that "passeth all understanding."

Bacon's political vision turns on new possibilities of desire. Even after a few days in Bensalem, the sailors are enraptured with this new world, "as was enough to make us forget all that was dear to us in our own countries." Their exhilaration shows how deeply the new eros has moved them, and how much Bacon hoped we would be moved in turn. Freud held that people functioning in society must sublimate their desires and suffer inevitable malaise. Unlike Freud, Bacon sees more possibilities for the erotic life than unsatisfying repression. He thinks that the new eros will be fascinating to its initiates; he anticipates that it will influence all of society in turn.

For desire to draw us in this new direction, Bacon considers that eros must be purified, heightened, and blessed. Even then, his optimistic vision is haunted by deep questions: can eros be so reshaped without dark consequences? Desire must be both chastened and heightened, avoiding paradox because this double motion reflects the complex repercussion of the new science on human sensibility. Bacon relies on what is most revered and "natural" to moderate the disorienting effects of altering nature. He also anticipates an alliance with Christian humanism, which shares the tenets of redemptive suffering and a new kind of love. This alliance would weaken if Christianity or classical learning were to decay. But Bacon cannot escape the dangers of Icarus. The pressing needs of suffering humanity call for these risky measures, but they are more than desperate gambles. In the end, Icarus is "kindred with heaven"; those who watch his wavering flight taste a strange exaltation. As Bacon notes, "there is no excellent beauty that hath not some strangeness in the proportion." Moved by beauty, along with hope and charity, humanity may take to heart a strange new science that "passeth understanding."

III

The Great Decryption

6

The Clue to the Labyrinth

But the universe to the eye of the human understanding is framed like a labyrinth, presenting as it does on every side so many ambiguities of way, such deceitful resemblances of objects and signs, natures so irregular in their lines and so knotted and entangled . . . Our steps must be guided by a clue. . . .
—Bacon, Preface to *The Great Instauration*

Even after the scientists have been tested and prepared, and after they wrestle with nature through experiment, the message they receive is still enigmatic. Nature's message is mysterious even when its text is disclosed. Before Bacon, others had begun to think of nature's secrets as if they were hidden in code. Bacon's contribution was to point out that the code may be solvable, something that his contemporaries doubted possible. He even set out a method of solving the code that parallels the techniques of codebreaking used in his time for diplomatic purposes. Though he did not anticipate the power of symbolic mathematics, by invoking the example of codebreaking, he prepared for the later union of mathematics with experimental science.

By the middle ages, thinkers were referring to the "Book of Nature," meaning that nature and the Book of Scripture both bear divine messages. However, the language in which the Book of Nature was written was hard to determine. In the earlier sources, it seemed that the Book was somehow in a natural, human language, though mysteriously expressed, since the creatures are not "readable" as words or characters. The Book of Scripture offered the prime candidate for the archetypal language. Scripture was the Book of Books, and its study was the touchstone for all interpretation of texts; its enigmas were much pondered, including the famous "writing on the wall" in the Book of Daniel, and pointed to the possibilities of secret writing. In fact, the Hebrew scriptures themselves display several ways to disguise words by substitution, which are among the earliest known ciphers.

Code and ciphers substitute certain symbols (the encrypted text) for the plain text. A cipher operates letter by letter, so that "Science" might be enciphered as "Uelgpeg"; codes work on larger groups of symbols, so "Science" may be encoded as "Sphinx." The few references to secret communications in ancient sources utilize only the simplest sorts of codes or secret writing; ancient writers thought it amazing that Caesar disguised his dispatches by substituting (say) *a* by E, *b* by F, and so forth, a simple cipher now used by children. In medieval times, literacy was so rare that any writing was effectively secret. Islamic-Spanish culture was the source for the later concept of cipher; the Arabic word *ṣifr* means *zero,* as in: "he's only a cipher." In late medieval France and Italy, there was a fashion for "emblems" or "devices," hieroglyphic images capable of representing a concept without using words (figure 6.1). Many Renaissance scholars tried their hand at these emblems, inventing new hieroglyphs on the model of the ancient Egyptians and thinking further about the possibilities of encoding messages in symbols.

Figure 6.1
Cesare Ripa, Emblem of Nature. Her turgid breasts denote the form "because it maintains created things"; the vulture on her hand denotes matter, "which being altered and moved by the form, destroys all corruptible bodies" (*Iconologia,* 1545).

Complex cryptography emerged fully during the late middle ages and Renaissance. By Bacon's time, diplomatic correspondence was routinely enciphered. The intercepted dispatches of foreign powers were opened and, at least in Italy, France, and England, subjected to attempts at cryptanalysis. This modern word denotes the solution of a ciphered text by unauthorized persons; the intended readers "decipher" it. The development of techniques of cryptanalysis was quite late. Its first efflorescence seems to have been in Venice, in the early years of the sixteenth century, but soon there were master codebreakers elsewhere in Italy and France. Those familiar with sophisticated ciphers began to associate them with the Book of Nature. In 1586, the French diplomat and cryptologist Blaise de Vigenère wrote that "all nature is merely a cipher and a secret writing," as if this were already a familiar concept. However, Vigenère assumed that stars or plants encrypted their divine message in a natural, human language, on the model of certain ciphers known to him that could hide a given text in musical scores, or in the positions of stars in the sky (figure 6.2).

Vigenère denied the possibility of solving the divine cipher through human artifice; not only is the cipher too difficult, but also men are too corrupt. He thought that only through the esoteric Hebrew kabbalah did God permit his elect access to his cipher by providing them with the divine alphabet. They must follow an ancient esoteric tradition strictly, not question or strike out anew. The chosen few may learn ancient mysteries, but must keep the secret from the profane world. Vigenère thought that codebreaking, even of human ciphers, was "a priceless cracking of the brain, and finally a quite inglorious labor." Despite his knowledge of the practical success of others, Vigenère was sure that codebreaking was futile in all but easy cases.

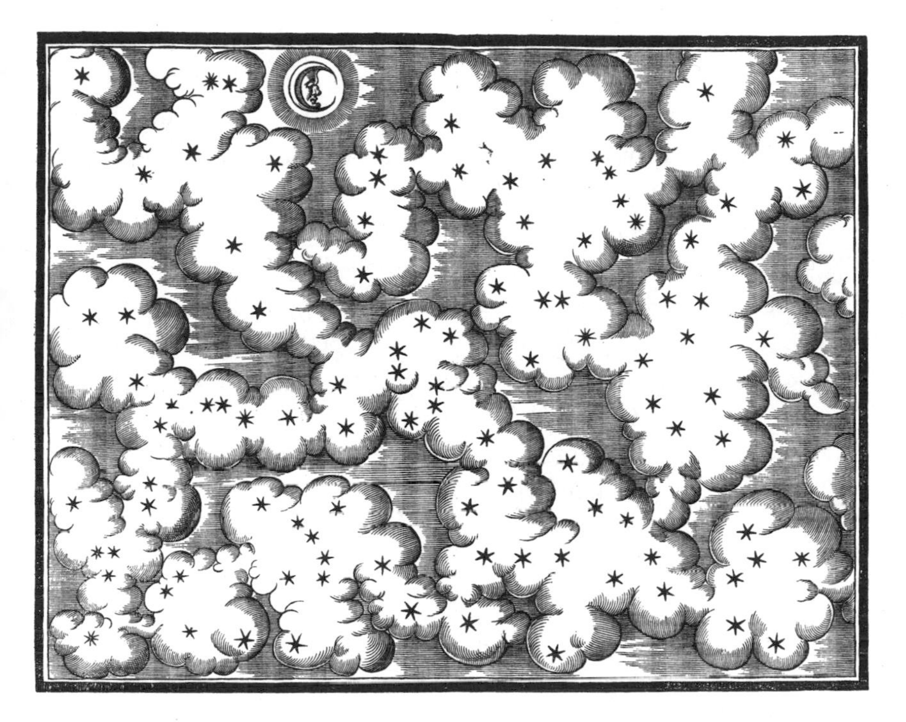

Figure 6.2
A cipher in the stars. The positions of the stars encrypt Psalm 19:1: "The heavens are telling the glory of God; and the firmament proclaims his handiwork." The cipher is made by dividing the picture into a "trellis" of 320 squares, in which stars can appear at a corner or at the center, indicating the letters by their positions (Blaise de Vigenère, *Tracté des Chiffres,* 1586).

In fact, there were already available powerful techniques of codebreaking that Vigenère pessimistically ignored, as the next chapter will discuss. Bacon showed considerable awareness of these new advances in the art of secret writing and had direct contact with its practitioners. His brother, Anthony, spent most of his time on the Continent, in charge of a number of "correspondents," who acted as spies abroad. In 1586, enciphered letters of Mary Queen of Scots were intercepted and cryptanalyzed, revealing a plot to assassinate Elizabeth, incite a Catholic uprising in England, and bring Mary to the English throne. The revelation of the exact texts of these letters gave evidence of Mary's complicity and led to her execution in 1587. Cryptanalysis also gave valuable evidence concerning Spanish designs on the English crown.

No one privy to these great events could miss the significance and utility of cryptography, and it surely was not lost on Francis Bacon, then twenty-six and at the beginning of his parliamentary career. Bacon spoke on the "Great Cause" of Mary's case as it was deliberated in Parliament. Through Anthony he must have gained further insight into the use of ciphers in more hidden affairs of state. Though Bacon constantly points to the reserved nature of things, he does so in order to *disclose* those secrets, always excepting divine matters and the secrets of the human heart. As Lord Chancellor, Bacon continually had to decipher King James's hidden intent, whether expressed in cryptic comments or the elaborate symbolism of the court masques the king cultivated.

Bacon considers ciphers central to the "Art of Transmission," the general study of discourse and writing. His interest goes beyond common letters and languages; he is interested in Chinese characters, as well as in the sign language of the deaf. Chinese he describes as being formed of "real characters," which represent things themselves, as do Egyptian hieroglyphics and gestures. These early forms of "transmission" unlocked important possibilities. An ancient tyrant once sent a messenger to a neighboring king; the messenger was to greet the king and then cut down the highest flowers in his garden. The gestures bewildered the messenger, but the cruel message was faithfully conveyed: kill the nobles. This "transitory Hieroglyph" fulfills some of the requirements of secret communications that Bacon lays down, "that they be easy and not laborious to write; that they be safe, and impossible to be deciphered; and lastly that they be, if possible, such as not to raise suspicion."

Bacon notes even more powerful ways to transmit and hide a message. As an example, he describes a cipher he devised as a young man, the single discovery for which he claims personal credit. Bacon is proud of this work, "for it has the perfection of a cipher, which is to make anything signify anything." That is, his cipher will hide *any* plain text in *any* cover text, given only that the encrypted text is at least five times longer than the plain. Bacon's biliteral cipher involves two stages. In the first, he represents the letters of the alphabet in terms of two letters only, arranged in groups of five (see table 6.1). Instead of *a* and *b*, one could have used any two recognizably different

Table 6.1
Example of an alphabet in two letters

A	B	C	D	E	F	G	H
aaaaa	aaaab	aaaba	aaabb	aabaa	aabab	aabba	aabbb
I	K	L	M	N	O	P	Q
abaaa	abaab	ababa	ababb	abbaa	abbab	abbba	abbbb
R	S	T	V	W	X	Y	Z
baaaa	baaab	baaba	baabb	babaa	babab	babba	babbb

symbols; two different symbols transposed through five different places yield $2 \times 2 \times 2 \times 2 \times 2 = 32$ different possibilities, enough to include all 24 letters of the English alphabet. Bacon notes that the "differences" need not even be between letters; instead of the letters *a* and *b,* one could use two bells of different pitch, or two different gunshots or torch signals. Thus this code is able to remove the letters from language and yet can reconstitute its meaning for the recipient. This feature, which it shares with gestures and "real characters," already renders it more fit to be a code of *things* rather than of *words* alone; it may be a step toward the symbolic code of nature, even though it is invented to serve the conventions of human communication. This code is also *binary,* like the code widely used for the machine language of computers.

Bacon then embeds this code into a cover text, probably because a stream of "differences," of gunshots or patterned bell ringing, might excite suspicion, or might be too difficult to convey with sufficient accuracy. Thus Bacon reduces the plain text FLY using the two-letter alphabet (see table 6.2). He then uses two different "forms" (fonts) of the alphabet, an "a" and a "b" form, similar enough so as not to arouse suspicion and yet different enough to register the *difference* between the forms (see table 6.3). Using these two very slightly different alphabets Bacon can then encode the plain text "FLY" into the innocent cover text, "Do not go till I come" (table 6.4). His example relishes the code's ability to convey just the exact opposite of

Table 6.2
Example of reduction

F	L	Y
aabab	ababa	babba

Table 6.3
Example of an alphabet in two forms

a form:	a	b	c	d	e	f	g	h	i	k	l	m
b form:	a	b	c	d	e	f	g	h	i	k	l	m
a form:	n	o	p	q	r	s	t	v	w	x	y	z
b form:	n	o	p	q	r	s	t	v	w	x	y	z

Table 6.4
Example of adaptation

F	L	Y
aa bab	ab aba	b a bba
do not	go til	l i com

the plain text in its cover text, not only avoiding suspicion but even misleading the enemy by a text that can mean exactly the opposite of what it seems. Bacon indicates that, if the ancients had used his cipher, they could have escaped detection and fooled their enemies. Despite its cleverness, Bacon's cipher does not seem to have had any actual usage in the diplomatic practice of his own time, which relied on far simpler devices. His cipher violates his own first rule, that it "be easy and not laborious to write," for minute variations of typography must be carefully guarded in order to render accurately the two different alphabets required. Bacon seems aware of such objections, for he complains of "the rawness and unskilfulness of secretaries and clerks in the courts of kings," on account of whom "the greatest matters are commonly trusted to weak and futile ciphers." He evidently had learned of this practical limitation on cryptographic ingenuity either the hard way himself or through the testimony of others. His publication of the biliteral cipher was a call for greater cryptographic security through stronger ciphers and more skillful clerks. Because of these practical limitations, such improvements were not implemented until several centuries later.

Bacon's interest in ciphers goes beyond their purely diplomatic use. His cipher suggests a more keen and suspicious reading not only of any given text, but also of any system of "differences," such as Bacon says may be found in anything seen or heard—that is, anywhere in Nature. This larger goal eclipses the narrower one of improving diplomatic ciphers, for which *concealing* a new sort of cipher would be the logical step, rather than publishing it. Bacon's readings of ancient fables are really "decodings." The outward symbols are "a veil, as it were, of fables," concealing hidden depths. He remarks that "religion delights in such veils and shadows, and to take them away would be almost to interdict all communion between divinity and humanity," so deeply are they required for the communion of such diverse and unequal minds. Parables were meant originally not to conceal the meaning, but to make it better understood. For instance, each detail of the Sphinx has a precise analogy with some aspect of Science. Likewise, nature's enigmatic messages are not perversely obscure; rather, we are too dull and impatient to solve them. His writings teach us an art of interpretation through decoding.

Bacon did not presume that the book of nature is written in any human language, even veiled by cipher. His crucial insight is that the methodical decryption of nature must grapple with a "language" that requires a whole new order of interpretation. Indeed, Bacon's yoking of *interpretation* with *nature* represents a great shift in understanding. For Aristotle, it is not nature that needs interpretation, but human language. For Bacon, "interpretation is the true and natural work of the mind when freed from impediments." Those impediments include the radical flaws in human understanding, beset with idols, and the labyrinthine intricacy of the world itself. The ancient labyrinth concealed the Minotaur that devoured Athenian children until Theseus killed it and escaped by means of a thread, or "clue." Bacon, the author of *The Clue to the Maze,* presents himself as a new Theseus, delivering humanity from the Minotaur of death by scientific secrets wrested from the labyrinth.

Just as Daedalus both built the labyrinth and found its deciphering clue, Bacon calls on science to breach the inmost sanctuary. Salomon's House is prepared to penetrate the veil over nature through a certain kind of interpretation whose key Bacon calls "induction." To find this key, Bacon envisaged a symbolic and schematic "alphabet of Nature." This key is not a fixed structure like a skeleton key, but rather a "key" in the cryptographic sense: a flexible indicator that guides decryption by delineating the structure of the cipher as it emerges. The essential preparation for induction is the exhaustive preparation of "tables and arrangements of instances, in such a method and order that the understanding may be able to deal with them"; Bacon also organizes his "alphabet of Nature" in similar tables. He cannot give full examples; however, he does give an extended attempt at tables regarding the nature of heat, leading to results strikingly like the modern view, in which heat is a form of atomic motion. What is important here are the tables themselves, which are manifold and detailed, going through many possible permutations of the instances, enumerating instances of "essence and presence" or "proximity where the nature of heat is absent" or "exclusion" or "degrees" of heat. These tables resemble the tables used for encipherment and decipherment, though applied here not to a natural language, but to "things themselves." A cryptanalyst examines the possible correlations between the appearances of certain letters in the cipher text, singly or by pairs or triplets, arranging the results in tabular form. Read negatively, this table also shows which ciphered letters are *not* correlated with which others. Other tables note the order in which letters are correlated, preceding or following others. Likewise, Bacon's tables marshal parallel data for heat, citing all known correlations and exclusions.

From the earliest sources on, cryptography had relied on such tabular arrays to give the visible key for the encipherment and, later, decryption. Given Bacon's detailed knowledge, it seems very likely that either he himself tried his hand at cryptanalysis, saw such work in progress, or heard accounts of it. His posing of a new, more secure cipher shows that he was fully aware of the powers of expert cryptanalysts and, quite likely, of their detailed methods. He certainly sets out his own biliteral cipher in tabular form. The cryptanalyst's tables are the necessary starting point to break the cipher in a systematic way. After that, certain deeply embedded linguistic features (such as the frequency of the letter "e" in English) can be much more readily brought to light and form the opening wedge to full solution. Bacon's tables proceed by the same logical categories of inclusion and exclusion, of quantity and correlation, that give the cryptanalyst's tables their revelatory power. In both cases, the tables are the beginning of *reading,* with all the interpretative acuity that word suggests. Telling passages in the cipher text must be located and probed; hypotheses need to be formed and tested, even if finally discarded, in order that correct order might finally emerge.

Both cryptanalysis and Bacon's hunt for the inner forms of nature require imaginative leaps that go beyond merely pedestrian accumulation of data. Although Bacon calls for a science that should be done "as if by machinery," he is also aware of the importance of the exceptional individual. The Fathers of Salomon's House are few, yet they rely on many others for the immense work of collecting instances. Only select minds can make the leap from the tables to the unifying insight that completes the work of discovery. Bacon gives images of these singular discoverers in his mythical tales; the solvers of the labyrinth are not to be confounded with their crews and companions, however valuable. The key to the labyrinth is a delicate thread that must be handled with care, for it might break. Bacon thought that true reading of the Book of Nature was reserved to those few who could follow the thread; the rest of humanity—including the wise king of Bensalem—must wait.

Moreover, even the scientists cannot anticipate what is to come; the great discoveries of the past had been quite unexpected. Genuinely new interpretations must "seem harsh and out of tune, much as the mysteries of faith do." He was wise enough to anticipate that he, too, would be surprised by what was to come. Experience can only be fully appreciated as it emerges, and not before. To read the Book of Nature means to experience it, to experiment with it, even at the risk of destroying old certainties. He envisoned a bridge between symbols, the "alphabet of nature," and things themselves. In so doing, he discerned a fundamentally new approach to the Book of Nature that transformed the character both of the Book and, finally, of Nature itself.

7

To Leave No Problem Unsolved

Putting together the mysteries of nature with the laws of mathematics, he dared to hope to be able to unlock the secrets of both with the same key.
—Epitaph of Descartes

Symbolic mathematics was the great missing element in Bacon's vision of the new science. Ironically, he included mathematics under the "inhuman" uses of torture, reducing "learning to certain empty and barren generalities; being but the very husks and shells of sciences, all the kernel being forced out and expulsed with the torture and press of the method. . . ." By "mathematics" he had in mind a sterile and rigid scheme of logical classifications, called "dichotomies" in his time. As he criticized the deficiencies of mathematics, though, he noted its unexplored potential. He might have thought differently had he known the very different character of the new directions being explored by some of his contemporaries. Though he may have known of potent techniques of codebreaking, he did not realize that similar methods could solve mathematical problems and could be applied to the Book of Nature. The flowering of symbolic mathematics brought the enterprise of decryption to a new level of power.

Here the stories of modern algebra and cryptanalysis converge, for they were both founded by the same man, François Viète (1540–1603). As a young man, Viète came into contact with the principal Calvinist leaders, among them the teenaged King Henry III of Navarre (later Henry IV of France), who developed confidence in him. During those years, Henry also befriended Anthony Bacon, on mission in France for the English and Protestant cause; it may well be that Anthony and Viète became acquainted. Viète went on to serve in high legal capacities and was given the title of counselor to the king. Though his intelligence and diplomatic ability recommended Viète for these positions, the few surviving anecdotes emphasize his absorption in mathematical matters. Sometimes he worked in his room for three days at a time without sleeping or eating. "Never was a man more born for mathematics," wrote a contemporary, who also noted that "Viète died young, for he killed himself through excessive studying." But Viète also possessed a rare and valuable skill that singled him out among other worthy advisers: he was able to solve the cryptograms of the Spanish and Italian courts, which were constantly attempting to intervene in the struggle in France between the Catholic League and the Protestant parties.

As a Protestant, Henry had to make good his claim to the throne through battle as well as diplomacy against the counterclaim of the League that no Protestant could be the legitimate king of France. His situation was precarious, for in 1589, the League held Paris and all the other large cities of France. Then, ciphered correspondence of Philip II, king of Spain, fell into his hands. Viète's solutions revealed not only the details of Spanish plans, but also the extent of the ambitions of the duke of Mayenne, the head of the League, who wished to become king of France—much to the surprise of the Spanish, who were scheming to put an infanta on the throne. Such inside knowledge was valuable as Henry worked out a compromise, in which he converted to Catholicism and Mayenne in turn submitted to Henry's authority. Viète evidenced a supreme confidence in his cryptanalytic powers. He reassured his king not to "get anxious that this will be an occasion for your enemies to change their ciphers and to remain more covert. They have changed and rechanged them, and nevertheless have been and always will be discovered in their tricks." Nor was Viète's confidence misplaced, for he continued to read Spanish and other dispatches right up to 1594, when Mayenne finally accepted Henry as his king.

When Philip realized that the French had been able to read the Spanish ciphers, he complained to the pope that they must have been using black magic to accomplish this feat. The pope ignored his accusation, doubtless aware that the papal cryptographers had been able to solve Spanish ciphers for thirty years without any diabolical aid. A contemporary chronicler noted that Philip's accusation only earned him the "contempt and indignation" of those who heard it. Spain seems not to have had cryptanalysts, which might explain Philip's attitude.

On his deathbed, Viète recorded his methods in a memoir to Henry's chief minister, Sully. Viète was aware that his techniques of decryption were not understood by most of his contemporaries and was worried that they would die with him. His memoir was secret and long remained unknown; the very existence of such methods of reading enemy dispatches was an important secret of state, and even more so the exact technique. It was no surprise, then, that rumors of such powers seemed incredible even to well-informed contemporaries, including Vigenère, who held that systematic methods of cryptanalysis were impossible. Even as late as 1771, Voltaire derided cryptanalysts for being "as great charlatans as those who boast of understanding a language which they have not even studied."

But Viète knew better. He recommends not only cultivating codebreakers of great native talent, but, even more, the implementation of "infallible rules" that make his methods accessible even to his skeptical contemporaries. Viète was the first to set down a truly methodical and general attack on ciphers, as opposed to informal reliance on lucky conjectures. Even though his memoir remained secret, he trained others in his methods. The "Black Chambers" of the following century incorporated such methods as the core of their institutional system of decryption.

During the same years that he was engaged in working out these methods of codebreaking, Viète was also at work recasting algebra in a new systematic form. To understand what he did, it is helpful to look back to Al-Khwārizmi, who lived in the eighth century A.D., and whose book transmitted the Arabic art to the West. *Al-jabr* means *restoration* or *making whole;* as in the addition of the same term to both sides of an equation, particularly in order to eliminate negative terms. The less familiar term *al-muqābalah* means *opposition* or *balancing,* as in cancelling equal terms on both sides of an equation.

Al-Khwārizmi enumerates simple problems with their solutions, but does not offer any general method. He does not use anything like the modern notation of algebra; there are no *x*s or equations, only statements in the rhetorical form of sentences about what he calls "roots," which now are called "the unknowns." Al-Khwārizmi groups his problems together according to the types of quantities that are being equated, such are squares equal to roots, or squares and roots equal to numbers, so there is some sense of the *types* of questions he can answer. However, in most cases, the solutions are so simple that no real method is expected or provided. For instance, in the case he calls "squares equal to roots" (for instance, $x^2 = 9$), he jumps immediately to the solution that the root is 3. His example carefully avoids any difficulty with irrational numbers by choosing a perfect square, 9. In the case of what we call quadratic equations, which he calls "squares and roots equal to numbers," Al-Khwārizmi provides a mysterious recipe for solving for the root. It is, in fact, equivalent to the general formula for the solution of a quadratic equation taught in high school algebra, but without the symbolic expression of its generality. Expressed in words it is literally a recipe, which begins: "You take first one half of the roots . . ." and continues through several subsequent steps. Its proof (like a cookbook recipe) seems to be all in the eating, for Al-Khwārizmi provides simple examples to show that it does work.

Such recipes parallel alchemical practices, which flowered in his time; both seek a valuable unknown, the gold, from the transformation and rearrangement of the given materials. Indeed, Al-Khwārizmi calls a *substance* (in Arabic, *mal*) what we call the "square" (x^2). This "substance" is prepared from "roots" mixed with numbers. Al-Khwārizmi is also aware that in certain cases the process does not work and "you have no equation," as when the square root of a negative quantity would have to be taken; he evidently does not recognize the possibility of the existence of such imaginary quantities, as they are now called. He is aware of the bareness of his recipes and feels obliged to "demonstrate geometrically the truth of the same problems which we have explained in numbers." He then presents a diagrammatic representation of what now would be called "completing the square," in order to demonstrate the truth of his recipe.

Thus, although the recipe might be used by someone who did not know or care about why it was true, Al-Khwārizmi feels that the recipe is rightly understood only by recourse to ancient geometry. In so doing, his practice is different from that of most of the alchemists, who are content to rely unquestioningly on the wisdom of the ancients, which they only seek to duplicate, and not alter, or even understand. Al-Khwārizmi's explanation comes only at the end, after the bare enumeration of his ingenious tricks. Lacking a general symbolism, he again presents it in a specific example. His diagram only illuminates the one case and does not really show us how to come up with other solutions to different types of problems from those he has discussed.

For a time, the practice of algebra was a rather obscure and undistinguished use of such tricks. In the sixteenth century, algebra came to be called the "art of the *coss*" (in Italian, *cosa,* the "thing" or unknown), sometimes the "great art," and was said to concern a hidden or occult quality of numbers. Exactly what algebra hides was not stated, but one infers that it may be matters not known to common acquaintance with numbers, and to powers attendant to such knowledge. It was also not clear how alarming or ambiguous such powers might be. The celebrated Elizabethan magus and astrologer John Dee considered algebra as part of arithmetic and praises its uses for merchants, travelers, navigators, mint masters, physicians, and all who must mix, graduate, or temper different substances, whether gold or medicine. Algebra is a potent and practical art, an important adjunct to what he calls "Archemaistrie" or "Experimental Science," which for Dee meant applied magic. One infers that algebra could also be of great use in "Thaumaturgike," the art of producing wonders. Despite Dee's respected position, in 1583 a mob broke into his home, damaged mathematical instruments, smashed optical apparatus (including a celebrated set of convex mirrors that the queen had particularly admired), and scattered his great library. It is not clear how much of this vandalism had to do with vague fears about sorcery or necromancy and how much to do with Dee's mathematical diagrams. Mathematics was widely suspected of being one of the black arts, and ordinary people thought it very dangerous.

Dee defends mathematical studies against the accusations of his "Brainsick, Rash, Spiteful, and Disdainful Countrymen." As recently as 1550, books that contained any type of mathematical diagrams were especially suspect and were burned as "Popish, diabolical, or both." If such obloquy attached to innocent geometrical diagrams, one can only imagine in what light the symbols of algebra were viewed. Although these panic fears may have been limited to the uneducated, algebra's reputation was low even among the learned. The "art of the coss" was considered suitable for public demonstrations, rather like magic tricks or acrobatic stunts, to awe and entertain the mob. As late as 1548, Ferrari and Tartaglia, both eminent Italian mathematicians, engaged in a turbulent public contest in Milan, in the aftermath of a scandalous controversy about the solution of cubic equations.

Despite the notable achievements of these Italians, algebra still lacked its characteristic symbolism and method; these were Viète's innovations, first described in his *Introduction to the Analytical Art* (1591). He compares alchemy to Arabic algebra, an art

> so old, so spoiled and defiled by the barbarians, that I considered it necessary, in order to introduce an entirely new form into it, to think out and publish a new vocabulary. . . . And yet underneath the Algebra or Almucabala which they lauded and called "the great art," all Mathematicians recognized that incomparable gold lay hidden, though they

used to find very little. . . . The metal which I bring forth yields the kind of gold which they wanted for so long a time. Either that gold is alchemical and faked or it is genuine and true.

Viète presents himself as a true alchemist, having fulfilled the dream of transmutation into gold by transmuting a defiled and barbaric art into something pure. René Descartes, somewhat later than Viète, also elaborated this new mathematics and expresses a similar asperity when he speaks of "the art which goes by the barbaric name of 'algebra'" as "a confused and obscure art which encumbers the mind, rather than a science which cultivates it." Viète depicts the Arabic algebraists toiling tediously and battling fire-breathing dragons, which may signify the difficulties encountered in solving a complex problem. The reason for their difficulties is that they "employed their logic on numbers," meaning that the Arabic algebraists had no general way of thinking. Even Diophantus, the great Greek mathematician on whom the Arabs drew, could solve problems only through devices tailored to the specific numbers in question. Diophantus's notation was a kind of shorthand for the words describing a problem, not the general symbols Viète and Descartes began to use.

Most of ancient mathematics was *synthetic,* proceeding from axioms to a theorem, as in the demonstrations of Euclid. But Viète emphasized that there was also "a certain way of seeking the truth" called *analytic,* which took the thing sought as granted and then worked backwards to find what would be needed in order to yield the stated conclusion. According to Viète, the ancients lacked "the science of right finding in mathematics." The analytic art is a *problem-solving art.* Viète cuts algebra loose from its traditional dependence on the "logic of numbers" by applying the techniques of arithmetic to what he calls *species,* whole classes or categories of numbers. To do so,

Viète introduces the term *coefficient,* which replaces the particular numbers that may be in a given problem with a general species; instead of the specific equation $9x^2 + 5x + 7$, he directs us to think of its general form $ax^2 + bx + c$. In Viète's "logistic of species," the true algebraic unknown is born in all its generality. Viète himself immediately notes that "this logistic is much more successful and powerful than the numerical one for comparing magnitudes with one another in equations." Through it, he has solved "the more famous problems which have hitherto been called irrational," including the trisection of an angle and the duplication of a cube, which baffled ancient geometers.

Viète claims that this art was known to certain ancient writers, but was since lost. He considers himself not an innovator but a restorer. How did Viète manage to find this hidden art of finding? The key may be the parallel between his mathematical and cryptological work. In both cases, he articulates a new kind of symbolism in which the cipher stands for the plain text (or the algebraic symbol stands for its implicit value). No less important, the search for the solution to the cipher (or for the unknown quantity) follows systematic procedures. Viète's symbolism of species is in essence a *cipher,* an amphibian concept that merges the general character of being a number with the particular numbers that satisfy the given stipulations. If the given equation is like the cipher text, then algebraic solution is analogous to breaking the code. To be sure, Viète's techniques for solving ciphers rely on the relative probabilities of finding different groups of two or three symbols together, and do not use algebraic equations directly. The *novel symbolic character* of algebraic symbol and cipher are closely akin, as is the equally important concept of *systematic solution.* The triumphant final sentence of his book proclaims that "the analytical art . . . appropriates to itself by right the proud problem of problems, which is: TO LEAVE NO PROBLEM UNSOLVED." Here the "proud problem of problems" is raised to the zenith of man's ambition.

Although Viète himself did not try to apply his new methods to solving the problems of nature, others soon did. As he announced his great discoveries, Galileo claimed that

> Philosophy is written in this grand book, the universe, which stands continually open to our gaze. But the book cannot be understood unless one first learns to comprehend the language in which it is composed. It is written in the language of mathematics, and its characters are triangles, circles, and other geometric figures without which it is humanly impossible to understand a single word of it; without these, one wanders about in a dark labyrinth.

The mysterious characters of the Book of Nature turned out to be mathematical symbols. However, Galileo read those symbols as geometrical figures, in the spirit of ancient geometry. His great contribution was to show that geometry could be applied to the problems of bodies in motion, to the physical world, and not only to mathematical propositions. Galileo demonstrated how idealized geometry lies behind the inevitable irregularities of the observable world; mathematics was the saving clue in the dark labyrinth of nature.

In the succeeding years, others augmented the symbolic powers of algebra and then turned them on nature. As a young man, Descartes abandoned the study of letters in favor of knowledge that "could be found in myself or else in the great book of the world." By interpreting the characters of the book as symbolic cryptograms, Descartes discerned a new path to the disclosure of its secrets. To explain his method, Descartes uses the example of reading something written in an unfamiliar cipher that lacks any apparent order. He notes that proper codebreaking uses system (rather than guesswork) to solve the secret writing. The implication is that cryptanalysis, though for Descartes a kind of game, is the paradigm of the systematic assault on the secrets of nature, penetrating always to the deeper truth of things by dint of its potent method. The realization of the full power of a generalized symbolic conception, whether of ciphers or algebra, was a crucial step in the development of modern scientific thought. Gottfried Leibniz, coinventor of the calculus, considered cryptology a crucial model for a universal science that would include a generalized, symbolic mathematics and promised to solve the cryptic text of nature.

The young Descartes journeyed about the world "masked as a soldier" (as he wrote in his private journal), a keen observer using the disguise of a combatant, the better to spy on the world. His chosen motto reflects his need to hide: *Bene vixit, bene qui latuit* (He lived well who hid well). Doubtless he was thinking of Galileo's confrontation with the Church, which alarmed many who were ready to follow the new directions in science. Descartes's mask hid him from political or religious danger, but also helped reveal him to those kindred spirits. He said that his seminal mathematical work, *La Geometrie* (1637), was written in an intentionally obscure style, as if to conceal his new techniques from all but the few mathematicians willing to follow him. At the same time, though, his publication of the new mathematics and its physical insights drew attention to those techniques, which had to be followed "as closely as he would the thread of Theseus, if he were to enter the Labyrinth." Descartes attracted the attention of those searching the maze while avoiding the gaze of unfriendly powers. A subtle call to arms, his reserved disclosures reflect and challenge the Book of Nature. The quest for hidden depths calls for secret guile, and with the advent of powerful mathematical methods, the time had come to begin the decryption of Nature.

IV

God's Spies

8

Kepler at the Bridge

With a pure mind I pray that we may be able to speak about the secrets of His plans according to the gracious will of the omniscient Creator, with the consent and according to the bidding of His intellect. . . . For these secrets are not of that kind whose research should be forbidden; rather they are set before our eyes like a mirror so that by examining them we observe to some extent the goodness and wisdom of the Creator.
—Kepler, *Epitome of Copernican Astronomy*

On a freezing day in 1610, Johannes Kepler was crossing the Charles Bridge in Prague and noticed some delicate crystals clinging to his coat. These snowflakes began a process of crystallization in his mind, mirroring the battle he had just waged to grasp the orbit of Mars. The planet and the snowflakes evoked Kepler's intense mixture of reason and ardor. The touchstone of mathematics tested and refined his vision; he glimpsed the outlines of a cosmic temple.

His first thought was graceful and poetic. He had been seeking a gift for a friend and patron, but had found nothing appropriate. His mind toyed increasingly with "nothingness," for his friend was "a lover of Nothing," as he playfully put it. Ironically, although Kepler was the Imperial Mathematician to Rudolf II, his coffers were always empty, for his imperial employer scarcely paid what he promised. Kepler wrote his little treatise *On the Six-Cornered Snowflake* (1611), thinking that his reflections, like snowflakes, were "the ideal New Year's gift for the devotee of Nothing, the very thing for a mathematician to give, who has Nothing and receives Nothing, since it comes down from heaven and looks like a star." Kepler warmed to his paradox, for the emptiness in his coffers sets off his mathematical "nothing," punning on the Low German *nix* (for *nichts*), which in Latin means "snowflake." Kepler's calling as a mathematician included what we now would call astronomy and astrology. In the snowflake he saw a heavenly form alight on his shoulder, like a bird of paradise, briefly in flight through terrestrial darkness.

Kepler wonders not only at the perfection of its form, but even more at its sixfold symmetry, which it shares with hexagonal honeycombs. In contrast, he notes that trees and bushes generally unfold in a *five*-sided pattern. Here the forces of life are at work; Kepler imagines the tender bodies of young bees finding comfort in their hexagonal nests. Though he speaks sensitively about the "feathered blades" or "plumes" of snowflakes, Kepler knows that they are lifeless. Eliminating external factors, Kepler seeks some internal cause to explain their sixfold shape. Nothing occurs at random, without reason, not even the pattern of a snowflake. "There is then a formative faculty in the body of the Earth," Kepler ventures. His mind vaults to exalted speculations about spirits, God the Creator, and solid geometry, but he mocks his own rapture as folly. Now questions clump around him densely, like the increasing snowfall. He tries many analogies, ranging from solid geometry to the strange patterns on the windows of Turkish baths, but these attempts only melt away.

Despite the inconclusiveness of these misshapen and inadequate attempts, Kepler returns to his "battle," not succeeding, but exhilarated. Though he relies on reason, Kepler's mind is not cool. He is animated, even feverish; he notices that, as each of his thoughts melts away, "this in its turn implies something that calls for more surprise than the very problem it was to have solved." He is spurred on even in his confusion, sensing "the chance of coaxing the truth from the comparison of many false trails." The labyrinth must have a center.

Besides many lumpish flakes, a few are perfectly six-sided. Suddenly he notices that those exceptional snowflakes are not just six-sided, but also *flat*. Why flat? Though he does not know, he is certain that there is a "beautiful aptness of *this* shape to *this* battle." He is unable to bring these clues fully to fruition, but knows that they *are* clues, that they may well lead somewhere, even though in his hands they still are nothing. Kepler notices that other crystals appear in the five figures of solid geometry that are regular (having identical sides) and convex (always curving outward). They are called the "Platonic solids," since tradition credited Plato with discovering that these were the only regular convex solids (figure 8.1). Kepler recognizes a connection between earthly shapes and heavenly forms. "But the formative faculty of Earth does not take to her heart only one shape; she knows and is practiced in the whole of geometry." Kepler notices a divine preoccupation with geometry, whether bodied forth in crystals or the perfect snowflake on his coat. Though geometry is literally "no-thing," not an object or palpable substance, it shapes the visible form of matter.

At the heart of matter must lie something that is not material, which is "nothing," if matter is presumed to be "everything."

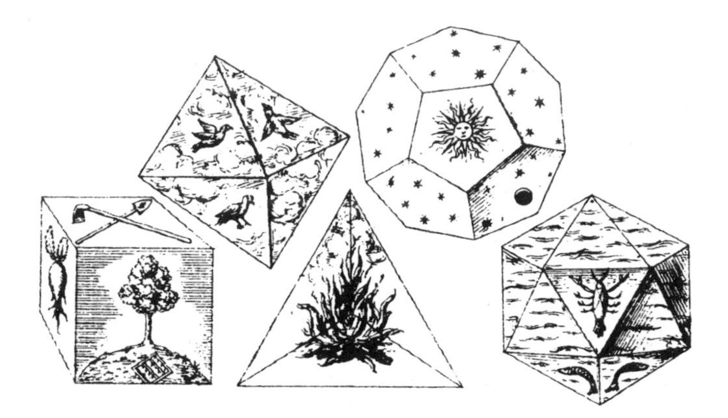

Figure 8.1
Kepler's depiction of the five Platonic solids, with the elements assigned to each: cube (earth), octahedron (air), tetrahedron (fire), dodecahedron (Sun and heavenly bodies), icosahedron (water) (*Harmonices mundi,* 1619).

He lays the matter at the door of the chemists and alchemists, thinking that they might be able to explain why the snowflake is six-sided. Kepler envisioned a bridge connecting the material, earthly world and the celestial realm of geometric knowledge. Led by the example of the snowflake on that cold bridge in Prague, he saw ordinary water take on a heavenly form, as if a star had alighted on his coat.

Crossing this bridge involved the use of mathematical arguments, which Kepler phrased in the classic geometry of Euclid. "Geometry is unique and eternal, a reflection from the mind of God. That mankind shares in it is because man is an image of God." Although Kepler was aware of Viète and his new algebra, he did not use it and expressed reservations on the few occasions he mentioned it. Kepler felt both attraction and skepticism toward the use of symbols in mathematics. Kepler remarked that

I too play with symbols; I have started a small work, *Geometrical Kabbalah;* it deals with the "ideas" of the things of nature in geometry. And yet, all the time I am playing I never forget that I am playing. For we can never prove anything with symbols; in the philosophy of nature no hidden things can be revealed by geometrical symbols, but only things already known can be put together; unless by sure reason it can be demonstrated that they are not merely symbolic but are descriptions of the ways in which [nature and the symbol] are connected and of the causes of this connection.

The work he mentions was never written or has disappeared, but perhaps he is referring to some sort of analytic geometry, such as Descartes later introduced. Whatever Kepler might have meant, there is no other trace in his works of a symbolic mathematics, and it was just in this direction that mathematical physics was to move. Yet his comment about symbolic mathematics anticipates a deep insight of Newton's: mathematics may reveal interconnections precisely because it veils the inner nature of things. Kepler also asserts the supreme importance of making a bridge between the symbols and physical reality, between mathematical theory and experimental data.

Accordingly, Kepler objected to contemporary attempts to draw on esoteric traditions and symbolic manipulations, for they kept mathematics from fruitful wrestling with quantity and experience. Robert Fludd, a prominent hermetist, scornfully asserted that Kepler's concern for quantity was a travesty of true philosophy: "For it is for the vulgar mathematicians to concern themselves with quantitative shadows; the alchemists and hermetic philosophers, however, comprehend the true core of the natural bodies; . . . [Kepler] has hold of the tail, I grasp the head; I perceive the first cause, he, its effects." For Fludd, true mathematics is "explained by means of hieroglyphic and exceedingly significant figures" that he calls "hieroglyphical figures." Like the "extracted essence" of alchemists, these hieroglyphs distill the secret "inner nature of the thing" and display it "as a precious gem set in a gold ring, in a figure best suited to its nature." (See figure 8.2.) Fludd continues to place the Earth at the center, not the Sun, and his arrangement does not try to render the precise distances of the planets, as Kepler does. The aim of such a hieroglyph is to draw the soul of an adept more deeply into the mystery that it symbolizes, interpretable only spiritually, not quantitatively.

There is a vast difference between the glyph of the mystic and the mathematician's diagram. Kepler retorts to Fludd, "*I hold the tail* but I hold it in my hand; *you may grasp the head mentally,* though only, I fear, in your dreams." Kepler flatly said that he hated all cabalists and objected to their insistence on secrecy and esotericism. He was certain that Nature's secrets ought to be brought to light. "God wanted to have us recognize these laws when He created us in His image, so that we should share in His own thoughts." In the prayer with which this chapter began, Kepler admits that there may be kinds of knowledge that are forbidden. Along with Bacon, he asserts that the knowledge of nature is not of that dangerous kind. He adds a mathematical reason: "For what remains in the minds of humans other than numbers and sizes? These alone do we grasp in the proper manner and, what is more, if piety permits one to say so, in doing so our knowledge is of the same kind as the divine, as far as we, at least in this mortal life, are able to comprehend something about these." The mathematical quality of this knowledge shows its divine provenance.

From early youth Kepler had sought to find the intimate relation between geometry and the visible world. His ideas seem to crystallize in a recurrent pattern, as does his religiosity, which saw in geometry a commentary on Genesis. The seed of this crystallization was Plato's vision that the creator impressed into

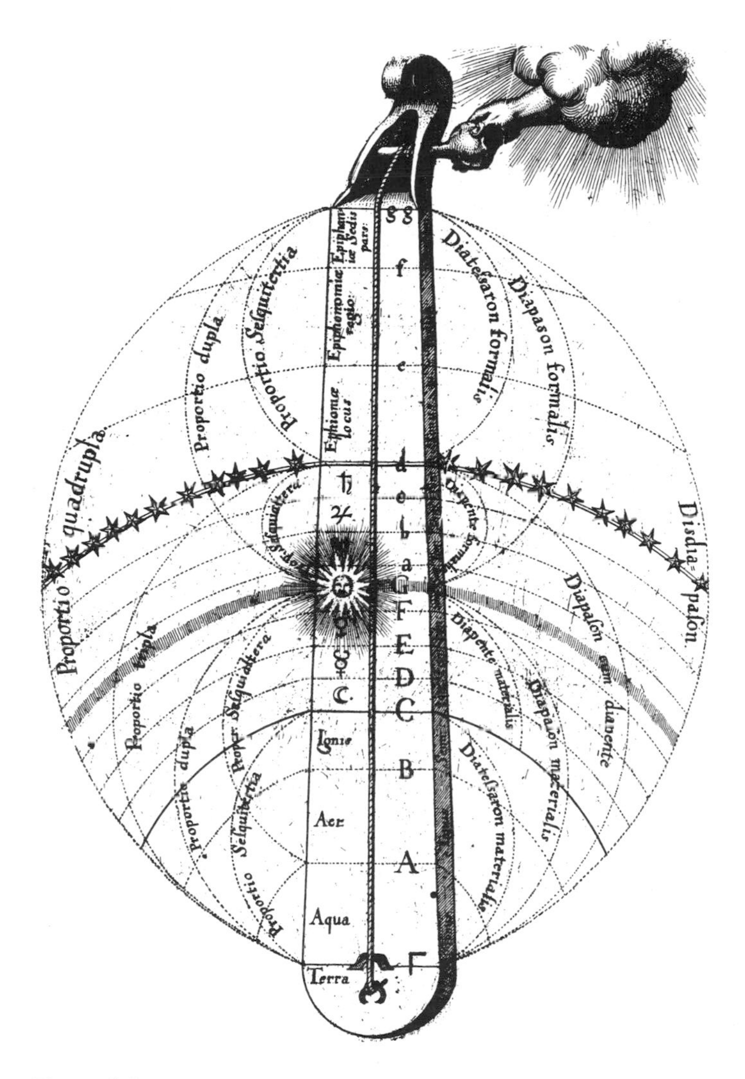

Figure 8.2
Robert Fludd, *Monochordus mundanus*. The hand of God tunes the harmony of the world (*Utriusque Cosmi . . . historia*, 1600).

the world the five regular solids mentioned above. This vision moved Kepler's first great speculative leap, which he noted down on July 19, 1595, aged twenty-four; it was a day he never forgot. On that day, Kepler noticed a connection between Plato's vision and the Copernican hypothesis, uniting God's shaping power as central Sun and as divine geometer. Kepler realized that the measured radii of the orbits of the five planets are close to the radii required when the five Platonic solids are nested together in a certain order (figure 8.3). This idea overwhelmed him with its beauty; he wept. He felt he glimpsed divine mysteries, despite his unworthiness. Kepler called his discovery "the secret of the cosmos" (*Mysterium Cosmographicum,* 1596). Looking back later, Kepler said that "the whole scheme of my life, studies, and works arose from this one little book." It also convinced him to become a mathematician rather than a theologian, as he had previously intended: "See how God is also praised in my work in astronomy."

With these divine secrets, Kepler entered "the presence of Zeus himself" and took his fill of divine ambrosia. He had been permitted to learn the thoughts of the creator as he formed the world, "the nature of the cosmos, God's motive and plan for creating it, God's source for the numbers, the law for so great a mass, the reason why there are six orbits. . . ." These are the most secret things in the whole universe, for "nothing in the nature of things is or has been more closely concealed, . . . nothing is more precious, nothing more splendid than this in the brilliant temple of God." As Kepler depicts it, the geometric figures of the planets form a three-dimensional labyrinth, traversing a vast volume, not just a flat plane. At the center of the labyrinth, the Sun radiates a celestial crystal.

Yet Kepler's delight did not blind him to the flaws in his youthful conception. The discrepancies between observations

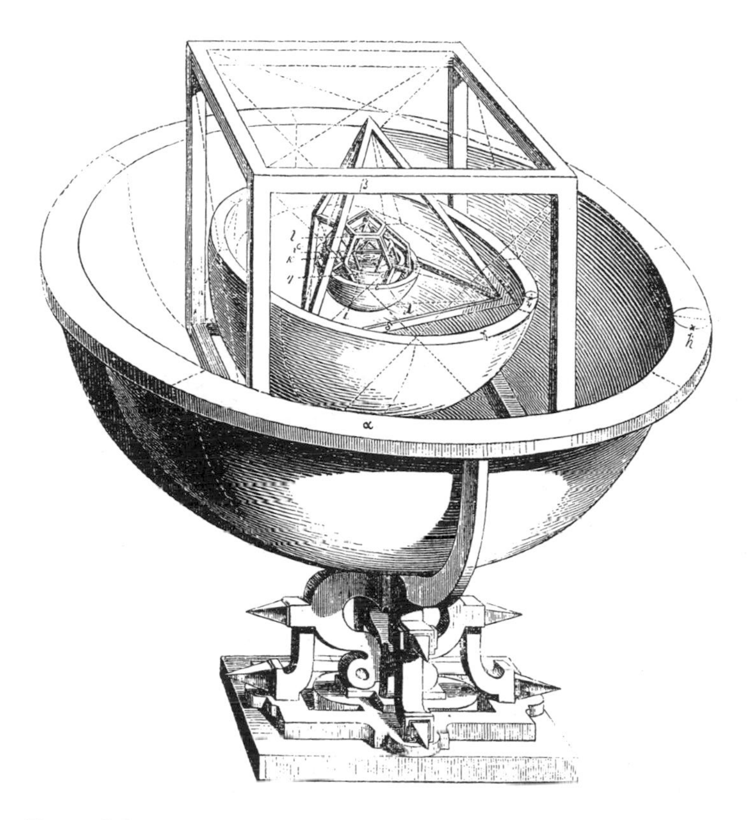

Figure 8.3
Kepler's nesting of the Platonic solids (*Mysterium Cosmographicum*, 1596).

and the predictions of his model piqued Kepler's curiosity; as his teacher, Maestlin, pointed out, the Copernican data was not sufficiently accurate to test the contemporary positions of the planets. When Kepler went to work with Tycho Brahe, he could get more precise data that would correct and perfect his model. For his part, Tycho may have been more attracted by Kepler's evident ability than by his questionable hypothesis.

These questions did not lead Kepler to disclaim his idea, which he returned to many times throughout his writings, aiming to perfect and complete it. Ironically, despite Kepler's confidence in it, later astronomers abandoned his vision of nested solids; on further investigation, planetary orbits do not conform to his geometric pattern. Still, modern astronomers have not explained the simple empirical formula relating the radii of the planets (the Titius-Bode Law). Someday a follower of Kepler's quest may find a convincing account of why the planets are spaced as they are. Kepler's question remains vital even though his answer failed. Yet, even if incorrect, this idea was the touchstone for Kepler's later work. This beautiful error enabled him to spy surprising truths, including his celebrated laws of planetary motion, so crucial for Newton's work.

First, Kepler set out to determine the orbits of the planets with great exactitude; then he sought to find a hidden explanation that would clarify what he had glimpsed of the *Mysterium*. This marriage of visionary intensity with keen attention to minute details set Kepler apart from arcane dreamers or from purely descriptive observers. It put him into an extended combat with the stubborn facts, in order to vindicate his vision through trial by battle. He considered himself "a disciple of Mars" not merely because so much of his life took place in the tragic shadow of religious war, but also because he understood that only "in a military manner" could he attain the depth of understanding he demanded. Kepler's brief "battle" with the snowflake is a miniature reflection of his far greater "battle" with the planet Mars, described in his earlier work, the *New Astronomy* (1609).

This struggle concerned the precise nature of the orbit of the planet Mars. In his dedicatory letter, Kepler brings before his imperial patron, Rudolf II, "a most Noble Captive, who has been taken for a long time now through a difficult and strenuous war waged by me under the auspices of Your Majesty." Throughout the letter, Kepler elaborates this mock-heroic metaphor of triumph that nonetheless celebrates a genuinely heroic achievement. According to Pliny, "Mars is the untrackable star," the ultimate challenge for an astronomer because of its irregular orbit. Also, Mars is of great importance in astrology; Kepler calls it "the Master of the Horoscope" and reminds Rudolf of the significance of Mars for the German nation. Some of Kepler's wit is playful, slyly begging on behalf of his ill-provisioned "army" of astronomers. Kepler's own salary was grossly in arrears and his pleading was based on urgent need. The "Martian metaphor" also gives a dramatic quality to a highly technical book. However, Kepler's style is not merely ornamental. It should not be forgotten that Kepler studied rhetoric extensively as a student, and for several years taught Virgil and rhetoric, as well as arithmetic, in Graz. This he did with distinction, as his superiors noted at the time. Kepler's works show a keen awareness of his audience and of the thoughtful practice of persuasion. He was quite frank in describing his labors in terms of battle and conquest. He confirms the truth of Bacon's images of wrestling, and the special heightening and tempering of scientific eros.

Turning back to the *New Astronomy,* one is struck by the impression the work makes of immense and confusing struggle. Kepler consciously chooses this confusing presentation in preference to the lucidity of Ptolemy, who sought "to lead [the reader] from self-evident beginnings to conclusions." Kepler also chooses not to give a mere description of the motions of Mars. Instead, Kepler decides to reveal his tumultuous methods, "in order to deserve and obtain the reader's assent, and to dispel any suspicion of cultivating novelty." In fact, Kepler's innovations are far more weighty and shocking than those of Copernicus, who remained faithful to ancient astronomy by adhering to circular orbits. Kepler's introduction of elliptical orbits represents an unprecedented departure. He emphasizes the complete *mathematical* equivalence of the theories of Ptolemy, Copernicus, and Tycho: "So far as astronomy, or the celestial appearances, are concerned, the three opinions are for practical purposes equivalent to a hair's breadth, and produce the same results." He is intensely aware that Copernicus worked from the same observational data as had Ptolemy, more than a thousand years earlier.

In contrast, Kepler works from new and much more accurate data he obtained from Tycho, who veiled all his observations in secrecy. Kepler had to sign a document promising to keep secret all information that he was given, though he deplored this practice as an "evil, shameful custom of the times"; he would have preferred to communicate his thoughts "to the masters of science so that by their hints I immediately make progress in our divine art." After his death, Tycho's family hounded Kepler to insure that their proprietary rights were not stolen. Despite (or perhaps because of) this careful secrecy, Tycho himself did not grasp the implications of his own observations, as Kepler later ferreted them out. Remaining faithful to the ancient circular pattern, Tycho revealed in his own work a new view of the world, yet could not bear to accept what he himself had disclosed. His crucial contribution was to measure the positions of celestial bodies with unparalleled accuracy. Ptolemy and Copernicus had claimed that their data were accurate to within about 10 minutes of arc (that is, to one-sixth of a degree, since 1 degree comprises 60 minutes). This was no small claim, considering that one's outstretched thumb held as far out as possible subtends 2 degrees of arc, or, equivalently, 120 minutes.

Nevertheless, Tycho found that their data were far less accurate than they claimed, sometimes more than a whole degree off. He was able to achieve the amazing accuracy of 2 minutes using only the naked eye, aided by large scales, and using several observers working together. For Kepler, at the critical point, accuracy of better than 8 minutes would be needed.

The case of Mars was crucial because its positions in 1593 were 5 degrees beyond the values predicted for them from the former data. This gross discrepancy was evidence that astronomy wanted reformation. It was partly happenstance that Kepler first began to work on Mars when he joined Tycho in Prague. Nevertheless, he realized that this chance boded some kind of "divine arrangement," for the motions of Mars "provided the only possible access to the hidden secrets of astronomy, without which we would remain forever ignorant of those secrets," for reasons to be mentioned shortly. This mention of the "hidden secrets" opens a special perspective on Kepler's singular rhetoric. Both Ptolemy and Copernicus characterized their theories as purely "mathematical." By entitling his book *New Astronomy Based on Causes, or Celestial Physics,* Kepler takes a very different position. He wishes to bring terrestrial questions of physics into the celestial realm, just as he saw in the snowflake a star come down to earth.

Ptolemy and Copernicus relied on circular motions, assuming that such motions were proper to the celestial realm. By interweaving earthly physics with heavenly astronomy, Kepler immediately calls that assumption into question and opens the way for his later finding that the planets move in the "aethereal air" with elliptical orbits. The only "physics" then known, that of Aristotle, was confined to the earthly realm of growth and change—*physis* in Greek—as opposed to the very different realm of changeless, heavenly order (*ouranos* in Greek). By altering the scope of physics to include the heavens, Kepler calls into question all known physics, as well as astronomy. It is not surprising that he adopts a new rhetoric to present his innovations. Kepler chooses to mingle what he calls "an historical presentation of my discoveries" with the more orderly and timeless exposition of the phenomena and the hypotheses that can account for them. Like his theories, his rhetoric also mingles earthly physics and heavenly bodies. Often Kepler simplifies and alters the strict record of how he proceeded; he leads the reader through much erroneous reasoning, which he later corrects. He compares the inclusion of these false or misleading trails with the wanderings of Columbus and Magellan, whose errors somehow give us "an enormous pleasure in reading." Kepler too gives us "the arguments, meanderings, or even chance occurrences" by which he first came upon his understanding.

This is partly to share the pleasure of his equally perilous journeys, but also has another goal. Kepler remarks that "although we by no means become Argonauts by reading of their exploits, the difficulties and thorns of *my* discoveries infest the very reading—a fate common to all mathematical books." Kepler calls on his readers to become Argonauts as they join the new Jason in quest of his golden fleece. By acknowledging ironically that readers of heroic quests usually do not thereby attain the status of heroes, Kepler nonetheless indicates that *his* readers would be subject to something like the perils that he himself surmounted. He would not deny them the "very great sense of pleasure" that he himself experienced "having overcome the difficulties of perception, and having placed before their eyes all at once this entire sequence of discoveries." When he says that this is "a fate common to all mathematical books," he means that the reader of such books must reexperience the demonstrations if they read the book adequately. His was not the way of the modern textbooks that strive to *hide* the difficulties of the theories they explain, preferring calculational ease to deeper understanding.

The "thorns" in Kepler's path are pricklier than those in the way of Ptolemy and Copernicus. They were not attempting to make heavenly phenomena *physically* explicable, just mathematically predictable; for Kepler the physical causes reach deeper, explaining why the world is as it is, and not otherwise. These hidden causes are the golden fleece of Kepler's quest, which requires passage through many deceptive paths. As in the journey of the Argonauts, the greatest heroes are forged in their trials, which also test and delight the reader. Kepler independently uses the same heroic imagery that Bacon invoked and makes explicit the difficulties that have to be surmounted. Kepler presents a synopsis to aid the reader and identifies it "as a thread leading through the labyrinth of this work," adding the image of Theseus searching the labyrinth to the heroic exertions of the Argonauts. Kepler admits that this guiding thread is incredibly complex, "more tangled than the Gordian Knot." His diagram of this synopsis spans eight full pages and gives visual form to this entanglement.

Kepler does not have the leisure to imagine all possible accounts of Mars and scrutinize them in advance of his struggle with the data. He does not approach the data naively, but always "theory-laden," bearing some interpretative hypothesis as a guide. This constant reliance on a hypothesis gives his work its labyrinthine quality. When Kepler calls up the actual path of Mars relative to the Earth taken at rest, he remarks jokingly that it looks like a pretzel (figure 8.4). Ptolemy never seems to have calculated this path, not considering that a heavenly body, like an earthly body, *could* have a path. Kepler refers

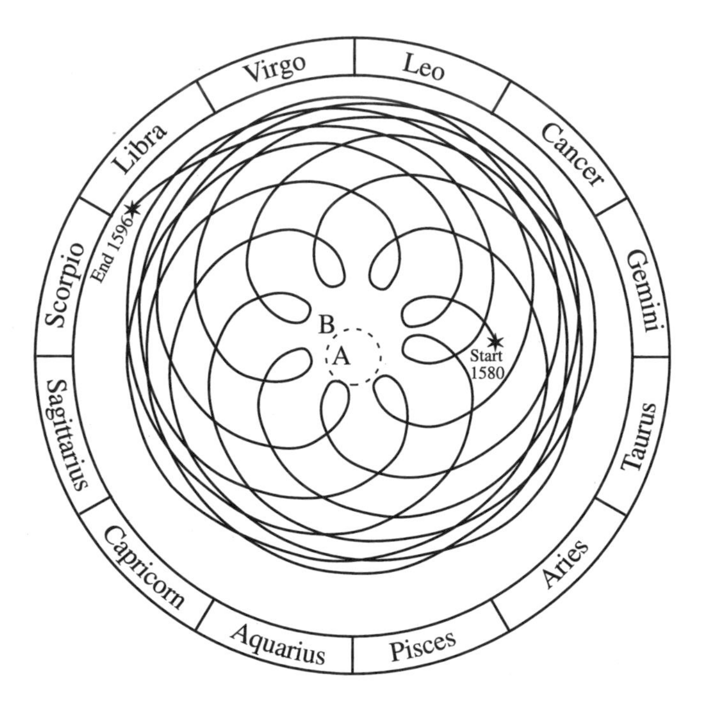

Figure 8.4
Kepler's diagram of the motion of Mars in the Earth-centered view (*Astronomia nova,* 1609).

to a "pathetic" contemporary attempt to give a diagrammatic representation of orbits on the Ptolemaic hypothesis as "a vast labyrinth of the most intricate curves, a human figment which the nature of things clearly disowns."

Kepler found himself hard-pressed to do better by constructing a hypothesis from physical reasoning. Though Copernicus had put the center of the Earth's orbit near, but not exactly at, the Sun, Kepler feels certain that the Sun should not simply sit enthroned in the middle, as if a passive spectator of the celestial motions. The Sun, like a visible God, must reach out to move the planets. Kepler drew his central physical postulate from Gilbert: the Sun stands at the center of the universe as a source of motive power, like a magnet emanating invisible force. He emphasized the decisive effect of this postulate on his subsequent work: "I erect the whole of astronomy on Copernicus's hypotheses about the universe, on Tycho Brahe's observations, and finally on the Englishman William Gilbert's science of magnetism." He even expressed the wish that he "had wings to carry me to England so that I might converse with him." Gilbert only went so far as to identify the Earth's magnetic power with its daily rotation about its axis, like one of his terrellae spinning near another magnet. Kepler extends this notion to the Sun, asserting that "the Sun is likewise a magnetic body" and will also rotate.

This rotating Sun emits what Kepler calls a *species*, here meaning form or perhaps emanation. Since this power reaches out from the Sun even to the furthest planets, it cannot depend on material substances that could diminish or be scattered. Kepler points out the resemblance between this and the behavior of light. "The remaining possibility, then, is that, just as light . . . is an immaterial *species* of that fire which is in the body of the sun, so this power which enfolds and bears the bodies of the planets, is an immaterial *species* residing in the sun itself, which is of inestimable strength, seeing that it is the primary agent of every motion in the universe." This hypothesis is for him the guiding thread, the mariner's compass, which guides him to the discovery of the elliptical orbits and the other laws of planetary motion. Based on this force emanated by the Sun, Kepler hypothesizes that Mars moves in a circle centered near the Sun (called an "eccentric"), though moving regularly about a *different* point (called the "equant"). The equant was Ptolemy's particular contribution to mathematical astronomy; it allows the geometric center of the orbit to be a different point than the center about which the planet moves *regularly.* Copernicus had rejected the equant as a disgusting artifice because it violated the postulate of perfect circular motion on which the theory was based. He detested the idea that the orbit could somehow have *two* centers, one of its shape and the other of its regular motion. Indeed, Copernicus took pride in having banished this monstrosity from his heliocentric astronomy.

Though Copernicus spurned it, the equant appealed to Kepler because it makes the planet appear to move more slowly the farther it is from the Sun, according qualitatively with his hypothesis about the active power of the sun, which is less strong when farther away. This "vicarious hypothesis" (as he calls it) is in some respects quite accurate; Kepler uses it to fit the orbit of Mars to within 2 minutes of arc. Explaining all the complex details of the iterative use of this hypothesis, Kepler notes that "if this wearisome method has filled you with loathing, it should more properly fill you with compassion for me, as I have gone through it at least seventy times at the expense of a great deal of time," for five years in all. This hypothesis, although false, nonetheless gave Kepler access to the truth; it was a stepping-stone to a better understanding. Kepler offers "a contrived thread, which nonetheless will lead you to the exit. If this method is difficult to grasp, the subject is much more difficult to investigate with no method at all."

Kepler uses the metaphor of the labyrinth quite closely; his use of the vicarious hypothesis is exactly parallel to the groping attempts one would try in a maze. One follows a hypothesis until a cul-de-sac bars the way. The thread records each trial, and so gives guidance. However, as Kepler says, the thread itself is "contrived," not yielding simple direction. It has to be "spun out" constantly, extending itself as the hypothesis ramifies. Indeed, the failure of a hypothesis gives more light than its success, for success may be illusory, but failure is clear. Kepler's vicarious hypothesis might have fooled a less wary seeker into thinking it was completely true. Though this hypothesis proved quite accurate in predicting Mars's longitude (measured from the Sun), Kepler nonetheless continues to test and vex it to determine where it might fail. Kepler realizes that it is, in fact, wrong because it does not give the correct distances from Mars to the Sun. If we had lived on the Sun, this could not have been realized, for it requires the "moving platform" of the Earth to measure it. Because the Earth is moving, we can infer this distance from measurements of Mars's position taken from different perspectives in the Earth's orbit, found at different times of year. The motion of the Earth does not demean its dignity, but enhances it, as if it were the miraculous vehicle the Deity assigned to give men the fullest view of the cosmos.

Careful comparison of the empirical center of Mars's orbit with its location according to the vicarious hypothesis led Kepler to a contradiction of 8 minutes of arc with Tycho's observations. Kepler immediately realizes this discrepancy to be decisive because it exceeds Tycho's stated precision of 2 minutes. Here he consciously stakes everything on the accuracy of Tycho's work. "Because they could not be ignored, these 8 minutes alone will have led the way to the reformation of all of astronomy, and have constituted the material for a great part of the present work." Thus Kepler speaks triumphantly even as "the hypothesis goes up in smoke," for its failure under the stress of interrogation convinces him that he must abandon at least one of its premises. Now Kepler realizes that he is "compelled to forsake the ancients and to search more diligently into these matters." He turns toward his new idea of celestial physics: the "species" emitted by the Sun pushes Mars with different force at different distances. However, the full magnitude of this departure only came home to Kepler slowly. Almost the entire text of the *New Astronomy* had been written before he realized that the orbit of Mars cannot be a true circle; this crucial departure from the ancients was, fittingly, arrived at in a particularly labyrinthine manner. Kepler is moved to astonished swearing as he recalls that "it must appear strange to an astronomer that there remains yet another impediment in the way of astronomy's triumph. And me, Christ!—I had triumphed for two full years." Kepler would not have realized this had he worked on the orbit of Venus or Jupiter; it was, he realizes, a stroke of divine providence that he studied a planet with a relatively large eccentricity (departure from circularity) in its orbit. Nevertheless, the orbit is so nearly circular that only Tycho's refined data could reveal it; even this "large" eccentricity only deviates about ½ percent from precise circularity.

Struggling with the "absurdity of the measurements" of Mars's orbit, Kepler seizes an odd image: "If one were to squeeze a fat-bellied sausage at its middle, he would squeeze and squash the ground meat, with which it is stuffed, outwards from the belly towards the two ends, emerging above and below from beneath his hand." Kepler calls his sausage-shaped orbit "oval," not at first identifying it mathematically as an ellipse. His homely image shows how he helped himself where his circular-oriented training failed him. As he grapples with his Proteus, it turns into a sausage. Laughter emerges from the labyrinth. His squashed sausage-oval explains why Mars seems to travel more and less swiftly at different points in its orbit by virtue of the varying distance of points in the orbit from the Sun. Kepler realizes that the "deformity" of the oval orbit is not a blemish; it expresses the precise nuance of Mars's orbit and purges his theory of its unwarranted assumptions.

Perhaps it was helpful that Kepler's own personal "orbit" deviated noticeably from the ordinary. Though a pious Lutheran, Kepler in his intense private reflection on theology was not fully in agreement with the Augsburg confession; though he differed with some Lutherans and Calvinists on the nature of Christ and of the sacraments, he sought unity among the warring churches. He was refused communion by his fellow Wertembergian Protestants for many years, a grave sign of his deviancy in their eyes. Kepler's own mother was prosecuted by other Protestants as a witch, though not for any doctrinal irregularity beside vague hysterical charges; only Kepler's energetic and skillful intercession saved her from torture and possibly death. Ironically, the Protestants were less tolerant of his Copernican views than the Roman Catholics. Though he professed himself a Catholic Christian, Kepler also was at odds with the Roman Catholics; he was expelled from their lands for refusal to submit to their doctrine. The intensity of Kepler's religious self-awareness may have helped him track Mars's orbit with similar attention; the principled deviation of man and planet are crucial, though subtle, and not to be neglected.

Along with his inner sense of Christian belief, Kepler felt a religious need to read the Book of Nature with fanatical care. In both cases, he is not content with what is generally believed because he senses a new message he may be the first to hear, and which has the urgency of divine revelation. "The very nature of things was setting out to reveal itself to men, through interpreters separated by a distance of centuries," the ancient astronomers and himself. "It is my pleasure to yield to the inspired frenzy, it is my pleasure to taunt mortal men with the candid acknowledgment that I am stealing the golden vessels of the Egyptians to build a tabernacle to my God from them, far, far away from the boundaries of Egypt." Rhetorically wrapping himself in the mantle of poetic frenzy, he identifies himself as an Israelite fleeing from idolatry and slavish opinions. "If you forgive me, I shall rejoice; if you are enraged with me, I shall bear it. See, I cast the die, and I write the book. Whether it is to be read by the people of the present or of the future makes no difference: let it await its reader for a hundred years, if God Himself has stood ready for six thousand years for one to study him."

This exaltation leads him to address his enigmas with utmost eagerness. Even though Kepler complains about tedium and frustration, his fiery determination is always clear. He also understands that this passionate intensity poses a threat to his ability to judge the implications of what he has done. Immediately after his realization of the oval orbit, he tries to grasp the physical cause of that oblateness, but later realizes that he had not proceeded thoughtfully enough. "I was blind from desire," Kepler confesses, as he prepares to explain the intricate web of his conjectures and errors. Having just escaped from one labyrinth, he immediately finds himself "entered into new labyrinths." The essence of his failure he attributes to his hastiness. As with Bacon's limping seeker, Kepler's testimony underlines the paradoxical danger that the scientist's intensity poses in the course of his quest. He must be avid, but also wary.

From one labyrinth Kepler plunges into another, as if his guiding thread had become more confused and labyrinthine itself. However, his climactic realization of the oval orbit encouraged him to plunge back into the melee. As a "contrivance," he summons back his vicarious hypothesis, which, now further clarified and tamed, continues to be helpful to him, for through its known properties he gains access to other information he needs. At a number of junctures in his work, Kepler devises and applies such hypotheses and notes ways in which their success might be merely delusive. He several times calls them *fictions.* At times, by using them, he even seems to be doing *worse,* since his successive attempts to improve the original circular vicarious hypothesis seem to be *less* accurate. He takes the reader into his confidence: "But, my good man, if I were concerned with results, I could have avoided all this work, being content with the vicarious hypothesis. Be it known, therefore, that these errors are going to be our path to the truth."

This striking claim brings to mind Bacon's devious-sounding maxim "*tell a lie and find a truth.*" As with the nested Platonic solids, Kepler's "lies" are not deceptions, but renewed wrestling. Not every hypothesis will avail equally in penetrating the labyrinth. Kepler's vicarious hypothesis provides accurate calculational help and provokes the crucial test of the circular orbit. At one moment Kepler proclaims "a triumph over the motions of Mars," but in the next moment, catastrophe strikes, "for while the enemy was in the house as a captive, and hence lightly esteemed, he burst all the chains of the equations and broke out of the prison of the tables." The distances measured around the whole circuit of the eccentric orbit Kepler calls "spies." They give him decisive information on the captive's wanderings, but nonetheless "have overthrown my entire supply of physical causes . . . and have shaken off their yoke, retaking their liberty." Kepler is almost reduced to desperation, but he rallies and gives orders to send "new reinforcements of physical reasoning in a hurry to the scattered troops and old stragglers, and, informed with all diligence, stick to the trail without delay in the direction whither the captive has fled."

Kepler does not shrink from exhibiting the vexation and ennui that must be endured. As he recounts one calculation, he breaks off to remark, "I can't imagine anyone reading this not being overcome by the tedium of it even in the reading. So the reader may well judge how much vexation we (my calculator and I) derived hence. . . ." At another crucial juncture, his physical causes "go up in smoke," leading Kepler to think all his work "futile." But it is at this very moment that he makes an important connection between two different passages in his work: he realizes that the number 429 is both a significant difference in the distance of the oval orbit from circularity, as well as a number he remembers from certain observational calculations. "It was as if I were awakened from sleep to see a new light," and once again Kepler finds his faith restored and his path made clear. In such minute details, Kepler found his great results, and he brings his reader through them with the utmost fervor. Although his "war" on Mars does finally lead to capture and triumph, Kepler's noble captive remains godlike. Kepler himself is "tormented" and "sweating," slaving away at his endless calculations. He makes himself the figure of fun, heightening rhetorically his mental dishevelment. As it dawns on him that the "oval" is really an ellipse, he tears his hair; at one point he says he was "considering and searching about almost to the point of insanity," but is still in the dark. When the light finally dawns, he exclaims: "O ridiculous me!"

Kepler's ironic humor expresses a fundamental enjoyment of his labors, or at least of the revelations those labors disclose. At one point he introduces a certain technical device "for fun," savoring a particularly sweet device. His groans over his immense calculations also are expressions of constant emotional involvement, almost as if they were cries of pleasure rather than of pain. Kepler makes clear that he is engaged not only in a battle, but in an erotic one at that, pursuing a "lusty wench" that seeks him mischievously, a character he calls Galatea, drawn from his reading of Virgil: "For the closer the approach to her, the more petulant her games become, and the more she again and again sneaks out of the seeker's grasp just when he is about to seize her through some circuitous route. Nevertheless, she never ceases to invite me to seize her, as though delighting in my mistakes." Kepler views his crescendo of errors and false paths as an extended arousal, whose agonizing delay turns pain into superlative pleasure. This unambiguous metaphor clarifies that his "seizing" is not a rape or violation, but rather the fulfillment of Nature's seductive teasing. Kepler's words can also be read to mean that the wench is throwing apples to attract his attention; perhaps Nature is dropping hints about the apple-shaped path ("puff-cheeked" as Kepler describes it) he will finally discard in favor of the elliptical orbit.

Kepler recognizes that his own meandering route may annoy some, but suspects that he will have a few choice readers who, like himself, "having overcome the difficulties of perception, and having placed before their eyes all at once this entire sequence of discoveries, will be inundated with a very great sense of pleasure." There is an irreducible mystery that keeps one from knowing the inner meaning of another's experience. Kepler's works are expressive of intense desires that he tried to share with his reader and that have a complex relation to individual passions. Whatever joys or sorrows he knew in married life, in his work Kepler mixes masculine and feminine imagery; the fluidity of his desire cannot be confined to conventional sexual roles any more than Mars's orbit conforms to ordinary preconceptions. Indeed, his treatment of the music of the planets includes both masculine and feminine sexuality, as he interprets the resolutions of major and minor thirds in the celestial chords. Kepler uses this rich mixture of imagery in order to engage us in a new domain of desire that is informed by the old, but not limited by it.

Here reason and ardor meet with an immediacy that might be called religious. For Kepler, mathematics was a mode of feeling, as well as of thinking. His vision delineated the proportions of that ancient temple, the cosmos, which Kepler calls "the corporeal image of God." The struggle to behold that mathematical image answers a divine call: "*Man, stretch thy reason hither, so that thou mayest comprehend these things.*" This desire may never be quenched, for there remains "that which is still completely unknown to us as well as that which we know and which amounts to only a small fraction of the other; for still more lies beyond." Here Kepler relied on a secret his geometric vision taught him, no less important to him than his famous laws, but far less well known. Plato held that the positions of heavenly bodies move in slow cycles; the sky we see tonight will recur in 26,000 years. But Kepler knew different. Because of the irrational relations between different geometric figures, the planets "will never return to the same starting point, even if they were to last infinite ages . . ." This sky will never return; an endless vista will follow. In the finite cosmos, Kepler discerned the signature of the infinite.

9

Newton on the Beach

I don't know what I may appear to the world, but, as to myself, I seem to have been only like a boy playing on the sea shore, and diverting myself in now and then finding a smoother pebble or a prettier shell than ordinary, whilst the great ocean of truth lay all undiscovered before me.

—Newton, to an unnamed companion

The new Theseus had returned from the center of the great labyrinth, bringing with him a rarely concordant system of the world. "No closer to the gods can any mortal rise," wrote the astronomer Edmund Halley, and the mathematician Pierre Laplace considered Newton's *Principia* the preeminent creation of the human mind. In 1802, the French visionary Saint-Simon even founded a "Religion of Newton." Newton's own reaction was complex. He surely knew the measure of his immense achievement, yet his self-depiction as a boy playing on the seashore is strikingly different from the triumphant "Sage and Monarch of the Age of Reason" venerated by his contemporaries. Aware of the great ocean of truth, Newton sensed the abyss between his mathematical thread and the secret he had sought at the heart of the labyrinth.

Where Kepler is expressive, Newton is impassive. In his public works, Newton avoids any image, analogy, or figure of persuasion, and achieves a different kind of deliberate power. Newton's model is ancient geometry; he adopts the spare style of definitions and propositions from Euclid. The title of his masterwork *Philosophiae Naturalis Principia Mathematica* (1687) emphasizes that he is presenting the *mathematical* principles of natural philosophy. By mathematics he means not only the application of quantity and proportion (which he shares with Kepler), but also their geometrical elaboration. Where Kepler rhetorically emphasizes his groping, Newton presents only his completed deductions. This accords with his role as founder of the city of modern physics. He is not the earliest settler, but the leader who ordained the constitution and design of the whole. The *Principia* established practically every branch of modern physics, which still rests on his foundations. Newton sets forth the elements of the calculus and applies them to bodies acted on by forces; he investigates bodies traveling through media of various sorts, as well as waves in sound or water. In Book III and "The System of the World," he infers the cosmic consequences, including the motions of the planets and their moons, as well as of the tides and comets. Though Newton does not address optics here, he did in a subsequent work; he mentions electrical and magnetic matters, but regrets that "there is not a sufficient number of experiments to determine and demonstrate accurately the laws governing the actions of this spirit." Faraday and Maxwell later followed his master plan "to discover the forces of nature from the phenomena of motions." Finally, Newton testifies that the lawful play of forces verifies the scriptural claims of God's dominion.

Writing in 1738, Voltaire noted that Newton "has discovered Truths; but he has searched for, and placed them in an

Abyss, into which it is necessary to descend, in order to bring them out, and to place them in full Light." To read Newton one must follow him into that abyss. His "Axioms, or Laws of Motion" are not "self-evident" starting points in the same way as Euclid's. These famous Laws have become so familiar it is difficult to regain the sense of their strangeness, yet puzzled students feel the difficulties that textbooks downplay. Solving the simplest problem of pulleys and weights requires a thoroughgoing mental adjustment, even in modern accounts that simplify Newton's methods. Mathematical points replace visible bodies; invisible forces represent palpable ropes or pulleys. Finally, the solution of a set of equations gives the bodies' positions, velocities, and accelerations. In response to their students' puzzlement, educators point out that most people still think of the world in Aristotelian terms, despite centuries of post-Newtonian physics (including subsequent, even more counterintuitive, physical theories). Whatever might be the true dimensions of such lack of education, ordinary intuition is at odds with Newton. Aristotle is far more commonsensical: moving bodies seem to come to rest, not continue moving indefinitely, as Newton's First Law implies. Newton reduces the world to invisible forces in a purified geometric continuum. He told his friend William Derham that "he designedly made his *Principia* abstruse." The original drafts of Book III of the *Principia* were written more informally in ordinary prose, rather than in their final form of austere propositions. Newton chose this final form specifically "to avoid being baited by little Smatterers in Mathematicks," as he told Derham. Only those who had carefully followed Newton's complex exposition could even begin to criticize. If they persisted, his unassailable demonstrations would defend him.

Newton's presentation aspires to the towering authority of ancient geometry. Though he was a master of the modern algebraic mode, he preferred the ancients.

> Indeed their method is more elegant by far than the Cartesian one. For [Descartes] achieved the result by an algebraic calculus which, when transposed into words (following the practice of the Ancients in their writings), would prove to be so tedious and entangled as to provoke nausea, nor might it be understood. But they accomplished it by certain simple propositions, judging that nothing written in a different style was worthy to be read, and in consequence concealing the analysis by which they found their constructions.

Like Viète, Newton thought that the ancients concealed algebra behind synthetic geometry, as he later did himself. Yet Newton argued that ancient geometric methods could solve the very problems that Descartes claimed were inimitable triumphs of modern algebra. Newton considered himself to have rediscovered the great truths that had been known to ancient sages. Like Bacon, he reinterpreted pagan and sacred legend in the light of his new project, even claiming that the inverse-square law of universal gravitation was hidden in the myth of Pythagoras's lyre. Meanwhile, in his voluminous private writings, Newton struggled to find the hidden sense of Biblical, alchemical, and hermetic texts. These private, even secret, studies shed light on his public projects.

The intense decryption that Newton applied to his enigmatic scriptural or alchemical texts also characterized his "reading" of Nature. For example, Newton's reading of Biblical prophecy shows precisely how those prophecies were fulfilled. Likewise, his system of the world shows how the force of gravity and the initial conditions of the planets lead to verifiable predictions. As with his secret studies, a certain reserved character haunts the most penetrating of his mathematical works. Newton had "the most fearful, cautious and suspicious temper that I ever knew," as Whiston, his successor in the Lucasian Professorship, remarked. Newton only granted the publication of his scientific works after extraordinary importunities and long delays. He wished to perfect those works, but also to hide them from prying eyes. This often-attested characteristic also fueled his notorious disputes over priority with Leibniz and others, disputes that could have been avoided by timely publication. Several times, Newton communicated a crucial discovery in a cipher, a secretive practice employed by other scientists, including Hooke and Leibniz. Besides drawing a veil of mystery over the discovery, the cipher also probed and tested its recipients, who must prove their mettle by decrypting it. By divulging his discovery in cipher, the scientist also magnified its glory.

Newton's esoteric preoccupations add further dimensions of secrecy. The practice of alchemy was by long tradition a secret activity, and Newton followed that tradition, studying carefully the original texts and adding his own comments in the same symbolic language. He was in contact with other adepts, including Robert Boyle, whom he worried might reveal powerful secrets to the unworthy, "not to be communicated without immense damage to the world if there should be any verity in the hermetic writers . . ." Newton's studies in Biblical interpretation and chronology were potentially no less explosive. Concealed in them was the secret of Newton's heresy, his adamant opposition to Trinitarian orthodoxy that he fortified by tenacious critical studies of early commentaries, as well as by studies concerning the possible corruption of crucial Gospel texts. As he well knew, such opinions would have ruined him during his life and, even after his death, were concealed for many years as a scandalous secret. These writings show the intense seriousness he attached to his precise theological position, which has variously been called Unitarian or Arian. It was a religion devoted to the rigors of the Old Testament, its Law and Prophets, and God the absolute Lord; Jesus was for Newton not "consubstantial with the Father," as the Nicene Creed has it, but subservient and obedient.

There are important connections between Newton's austere religious understanding and his scientific work. Newton's theological preoccupations were simultaneous with his most intense scientific work. Indeed, both Newton's original scientific and theological activity seem to have been terminated by his mental breakdown in 1693 (evidence of his intense struggle). Theology was not merely a recourse when he could no longer sustain his scientific work; in his old age he returned to both, but in a spirit of reviewing and correcting his earlier work. Though Newton never revealed his heretical opinions except to a few like-minded persons, he never specifically denied them. He allowed and even promoted the impression that he was orthodox by his careful silence and by his friendship with eminent churchmen; he even served on boards and commissions of the Church of England. However, his pointed avoidance of taking holy orders was doubtless noticed. He brought the box with all his secret manuscripts with him when he went to London after his breakdown. Newton did not destroy these potentially damning papers; even if he never looked at them again, they remained with him, enduring evidence of his hidden preoccupations.

The thread linking physics and religion seems to be finding the center of an enigma through gauging the significance even of the smallest phrase. In Newton's interpretation of a verse about the Last Judgment in the Book of Revelation, the phrase "night and day" stands out. From this Newton deduces the endurance of planetary motion and time itself even after the last trumpet. In Newton's "Rules for interpreting the words and language in Scripture" neither prophecy nor its interpretation was in any way rhapsodic or irrational. Among these self-imposed rules Newton includes "to assign but one meaning to one place of Scripture, unless it be by way of conjecture. . . . To keep as close as may be to the same sense of words, especially in the same vision." His work contains a series of symbolic identifications, identifying each recurrent symbol of the prophetic texts with one entity or group of people. Newton decoded the scriptures, rendering them concordant with recorded history. His exhaustive work went beyond common generalities (the hackneyed identification of the Whore of Babylon with the Pope) to the precise details of past and future happenings. Newton insisted on close reading of the symbolic images, rather than loose construction.

Newton also sought confirmation of the veracity of the prophetic writings and their hidden historical meaning. In his prefatory remarks, Newton advocates this study to "not all that call themselves Christians, but a remnant, a few scattered persons which God hath chosen." His tone is ominous; he invokes the disastrous example of those who did not understand the prophecies and rejected their Redeemer. The prophecies may save us, if we decode them, for they contain clues to the identity and activities of the Antichrist, who will lead many to perdition. He warns us that "this is no idle speculation, no matter of indifferency but a duty of the greatest moment." No hint of love or grace distracts Newton from his Herculean task. As he prepares to enter the labyrinth of prophetic symbols, Newton also warns us that they are a trial through which God tests and purifies us, ratifying our predestination. For Newton, rigorous prophetic interpretation is the path to salvation, separating the elect from the unworthy: ". . . the mystical scriptures were written to try us. Therefore beware that thou be not found wanting in this trial." This trial echoes Bacon's advocacy of heroic struggle with divine codes. The same sense informs Newton's experimental work in optics, particularly his celebrated *experimentum crucis,* the "crucial experiment." Newton himself gave this name to a decisive test of his theory that white light is actually compounded of many colors. Newton subjected light to a trial that also tried him.

Quite early on (probably in 1664 or 1665), Newton had experimented with prisms and observed the spectrum that first pointed him towards the radical supposition that all colors together compose white light. But the theory was open to criticism. Could not the prism itself somehow have modified the light, acting on the simple white light so as to force it to manifest a spectrum? Newton refined his original experiment to put this theory of modification to a stringent test. In what he called the *experimentum crucis,* Newton ingeniously redoubled his original scheme by adding a second prism that could be applied to the spectrum of light emergent from the first. He also masked the first prism so that he could isolate each color and test its refraction through the second prism (figure 9.1). All these modifications try the white light ever more severely; if white light is violently distorted by the first prism, the second prism should augment that distortion. On the contrary, Newton's apparatus showed that blue light emerging from the first prism is not modified further by the second prism; it remains blue, not breaking up into a second spectrum under the vexation of the second prism. Newton accepts the challenge of the modification theory, seeks to modify light even more drastically, and, failing to do so, gives powerful evidence that the colors really are intrinsic components of white light. Rather than being a violation, his experiment brings out the true nature of the colors and their integrity, which otherwise would have remained latent in the

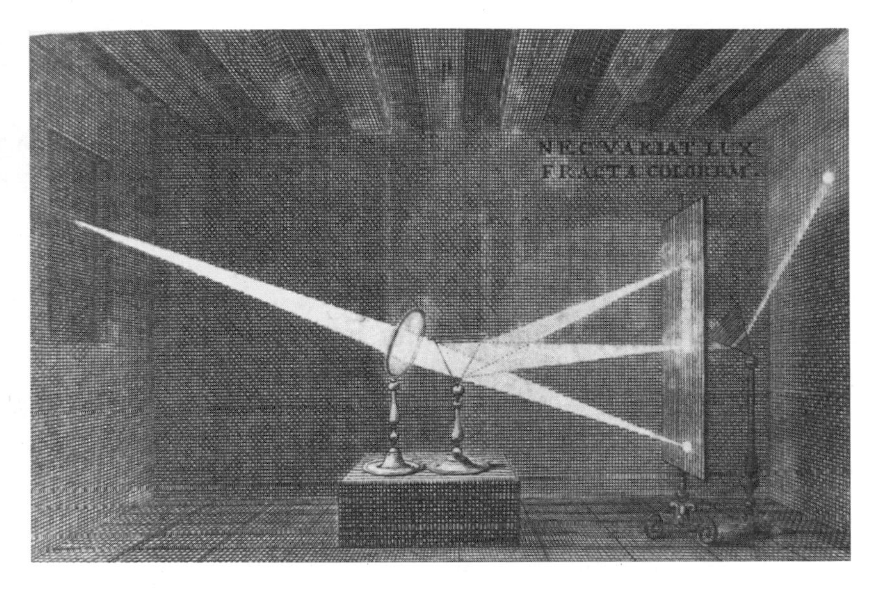

Figure 9.1
Newton's *experimentum crucis* (*Optique*, 1722).

white beam. Newton records a summary judgment from this trial: *Nec variat lux fracta colorem* (Light does not change color when it is refracted). This verdict proclaims the true nature of light, borne out in its trial by experiment.

Newton's phrase *experimentum crucis* originated with Bacon. As an important form of the "experiments of light," Bacon meant by this term instances of the crossroads (*crucis*) "where the roads part" and "the understanding is so balanced as to be uncertain to which of two or more natures the cause of the nature in question should be assigned." Bacon mentions some of Gilbert's experiments as aspiring to such decisiveness, although he criticizes them and offers improvements to make them more "decisive and judicial" in character. It took Newton from 1666 to 1672 to refine his initial idea so that it fully deserved being called a critical experiment. The perfection of such a test is itself the critical test of its contriver; Newton's optical theory was forged and perfected in the same crucible as the light it tested. This pattern of purification through intelligent ordeal also resembles the first part of the alchemical quest, *l'oeuvre au noir,* or "blackening", which Newton characterized as a "gross work," which subsequently "purges the matter from its gross feces & exalts it highly in virtue & then whitens it." In both alchemy and prophetic interpretation, the method and the goal at least of the beginning of the work is a radical purgation. By hewing to strict literality guided by rules, in his prophetic studies Newton required a precision of interpretation that is the analogue of the mathematical exactitude he enforced in natural philosophy.

This analogy points to what might be called an algebraic approach to prophetic interpretations: collation and analysis of texts solve the symbols in turn, each one pointing the way to the next. Then the prophetic symbols are checked for historical accuracy against external records. Though Newton felt he had adequately confirmed the prophecies' accord with history, he knew the pitfalls of relying on shaky data. He was doubtless aware of his intimate friend John Craig's *Theologiae Christianae Principia Mathematica* (The Mathematical Principles of Christian Theology, 1699), whose title echoes Newton's *Principia.* Craig used elaborate algebraic equations to demonstrate the truths of Christian teaching and to predict the end of the world, 1,500 years later. Newton was also aware of the mockery the French scholar Nicolas Fréret made of Craig's work, particularly of its presumption that such mathematical means were appropriately applied to apocalyptic matters. In comparison, Newton's own interpretative methods have a certain geometric, rather than algebraic, stance. He describes prophetic language as "Hieroglyphical" and compares it to Egyptian picture-writing. Newton agreed with the prophetic interpretations that Joseph Mede had pioneered before him, and particularly with the "synchronism" Mede introduced as a guiding tool of prophetic interpretation. As Newton phrased it, this implied that "the parts of Prophesy are like the separated parts of a Watch. They appear confused & must be compared & put together before they can be useful, & those parts are certainly to be put together which fit without straining." That is, a merely "algebraic" and chronological treatment of the texts misses certain recurrent features that are not sequential but synchronous, as if crucial elements of the prophecies were separated in different accounts. Mede and Newton sought to reassemble these separated passages and to restore the true shape of the prophecy. Newton chose the same image of a watch to describe the mechanical cosmology of planets.

During the years in which Newton prepared his mathematical theory of the world in a geometric form, he was engaged in an attempt to give visible form to the prophecies by deducing an accurate ground plan of Solomon's Temple (figure 9.2). Though it seems that the temple as such was not important in his interpretations, his project of reconstructing it in great detail is an apt example of this synchronic, geometric trend. In order to complete his design, Newton learned Hebrew and reconciled the various accounts in scripture, comparing the tabernacle of Moses with the temples of Solomon and Ezekiel. This included extensive study of the exact size of the "sacred cubit" to specify the true dimensions of the whole structure. After completing this immense work, Newton felt that he could gaze on the temple. He beholds the larger geometry of the divine edifice: "This structure commends itself by the utmost simplicity and harmony of proportions." Behind the temple and the cosmos

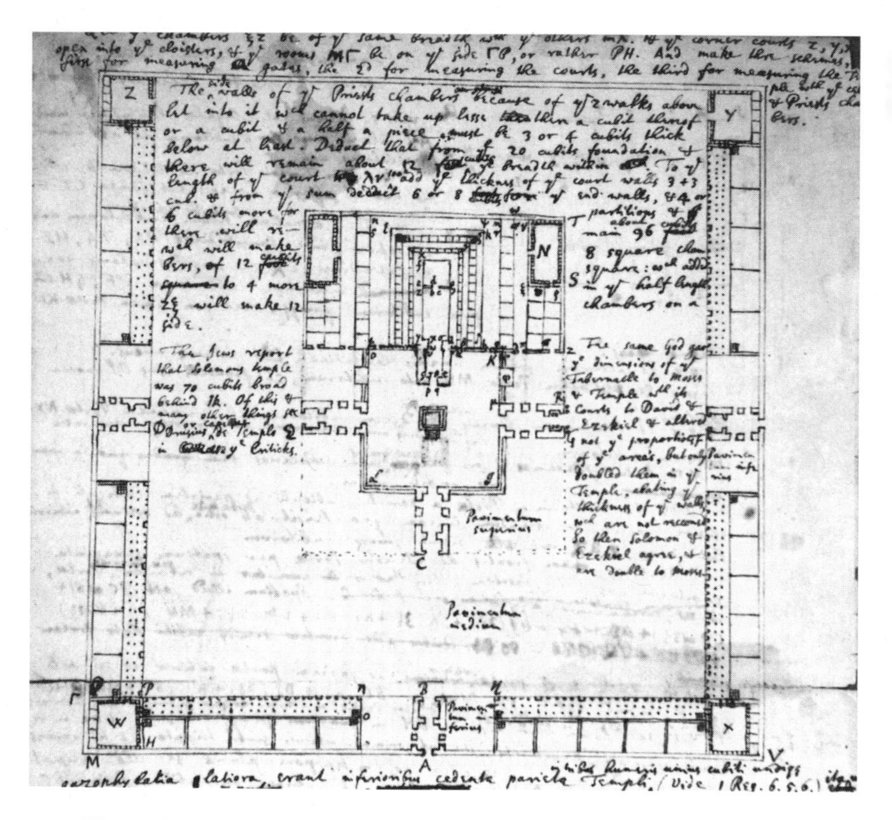

Figure 9.2
Newton's plan of the Jewish temple (Grace K. Babson Collection of the Works of Sir Isaac Newton, MS 434).

stands "Lord God *Pantokrator,* that is, universal ruler." Newton emphasizes the full extent and lawfulness of divine dominion, for "the lordship of a spiritual being constitutes a true god." His geometric treatment of physics delineates this dominion throughout the cosmos, bringing together space and time. Though he turned to ancient models, Newton considers curves that go beyond Euclid's straight lines and circles. He studies ratios of lines as they vanish, something that has few precedents in classical geometry.

Although Newton knew Viète's algebra, he relied on geometrical methods to present calculus in the *Principia.* For instance, he uses geometric tangents and chords to form an ingenious "microscope," which renders an enlarged image of the slope of a curve approaching a given point. Newton uses forces to measure the active principles that move passive matter in the course of time. He is aware of the difficulty of his ideas; his "microscope" glimpses the "evanescent" instantaneous velocity of a body "at the very instant when it arrives" at a place. He includes in the same diagram lines that show the spatial path of the body, its velocity, and the magnitude of the force it is experiencing. In contrast, Galileo had carefully avoided ever putting spatial and temporal magnitudes in the same geometric figure, for that would presume that space and time are the same sort of magnitude. Newton draws together space and time, giving visible shape to the vanishing moment. In his own way, he began the project of uniting space and time, which later Einstein would take up in another way.

Later, Newton articulated his algebraic version of calculus as "fluxions," meaning the "swiftness of growth" of some quantity, its velocity or rate of change obtained through an infinite series of ever-closer approximations. The invocation of the infinite, though powerful, can lead ordinary logic to paradox. Algebra must use infinite series, or appeal to infinitesimal magnitudes, and so Newton argues that "since this limit is certain and definite, the determining of it is properly a geometrical problem." He retained his original geometrical exposition unchanged in the later editions of the *Principia,* implicitly emphasizing its continuing superiority over algebraic presentations. Newton's geometrical treatment avoids many of the treacherous paradoxes that follow less rigorous algebraic approaches. Newton brings to the fore an even more fundamental limitation of algebra: no oval curve has an area that can be found "by means of equations finite in the number of their terms and dimensions." His simple geometric proof shows that *any* planetary orbit encloses an area that cannot be expressed by a finite number of algebraic terms. In modern terms, Newton had discovered the immense domain of the "transcendental numbers," magnitudes that are not the solution of any equation with a finite number of terms. It was early suspected that pi was transcendental (although it was not rigorously proven until 1882), but the greater surprise is that *most* numbers are transcendental; the algebraic numbers are densely surrounded by infinite swarms of transcendentals. Newton's proof shows that algebra is unable to express the most basic features of curves (their areas or, what is the same, the time a planet takes to complete an orbit) without recourse to infinite series. Geometry, by contrast, reaches immediately to those results, since its magnitudes are continuous, not restricted to solutions of algebraic equations. Thus Descartes's plan to reduce all physics and mathematics to algebraic equations cannot succeed unless the equations have an infinite number of terms, with all the problems and paradoxes such infinities invite. Newton's geometric diagrams cut through the labyrinthine complexities of these infinite series, as if geometry has wound up the thread that algebra must painfully trace.

Newton's realization about the way geometry transcends algebra was far ahead of its time. It is an insight whose significance is comparable with the ancient Pythagorean discovery of irrational magnitudes like the square root of 2. Beside this profound mathematical realization, Newton found in the calculus the crucial way in which motions might be correlated and made intelligible in terms of invisible forces. To achieve his triumphant account of the visible realm of motion, though, Newton consciously sets aside all questions of what lies beyond observable correlations. He abstracts from natural phenomena the mathematical form of the forces, and then uses those forces to explain other phenomena. In his public writings, he also sets aside the problem of explaining from first principles *why* those forces should have just those forms. He turns his attention away from the hidden depths, though he allows that a later generation might be able to go further and give some account. However, his project gains its power precisely by renouncing the need to give an account that goes beyond the mathematical. This renunciation may reflect the impossibility of any such account.

The hidden core of nature conceals itself at the moment its laws seem to be revealed. As the price for the mathematical description of nature, Newton had to abandon any claim to know the true causes of the forces that he invoked, saying *I do not feign hypotheses.* He may mean by "hypotheses" the *wrong* hypotheses, premature or mistaken, particularly Descartes's conjecture that the true mechanism of gravity was vast etherial vortices that propelled the planets in their orbits. Newton tried to demolish this popular hypothesis in the *Principia,* though his proofs were later shown to be mistaken. However, he noted that "force arises from some cause." Earlier, he had advanced several possible mechanisms for gravity, including one based on an interplanetary medium or ether, as well as a quite different one that ascribed gravity directly to divine causation. But, for various reasons, none of these conjectures satisfied him, so that his disclaimer at least indicates that he had left the question for later work, whether by himself or others. In the final paragraphs of the *Principia,* he asserts that "it is enough that gravity really exists and acts according to the laws that we have set forth and is sufficient to explain all the motions of the heavenly bodies and of our sea." This certainly is "enough," if the goal is to predict those motions with great accuracy. Newton's ending set the deeper questions aside as impertinent to the real project of prediction, which can go forward without any need to know the "cause."

By setting aside the inquiry into ultimate causes, Newton elevated the mathematical principles to the forefront of attention. This decision was prescient; Einstein's gravitational theory moves even further from such mechanical explanations. Despite his accomplishments, Newton felt that something lay beyond. His dreamy, even wistful, self-description as "a boy playing on the sea shore" seems to open a different perspective onto the triumphant discoverer. Newton recognizes some crucial respect in which his discoveries seem to him shells from the depths, not the great ocean itself. His description evokes an indefinable note of longing for the oceanic abyss. Newton repeatedly dismissed his mathematical works as "divertissements"; he also describes the boy as diverting himself. As successful as his mathematical principles had been, he had to pursue them by setting aside what he more deeply wanted to know, the profoundest secrets symbolized by the sea itself. His failure to plunge in was not ignorance, though the image of the boy suggests a kind of innocent disregard. Not only does Newton imply that he neglected the greater for the less, but his method required him to deal with the mathematical shells.

If one tries to unmask the forces and discover what lies behind their mathematical form, no matter what sort of interaction one proposes to explain the forces—whether bumping particles or ethereal swirls—that interaction will finally take the form of another force. Any mechanism explaining what causes a force will eventually have to say how the mechanism itself works, and that will involve talking about the forces that govern the meshing of the wheels or cogs. The concept of force renders mechanics intelligible, but it cannot look underneath itself. In view of this disturbing problem, when Newton asked about causes, he was not really asking for some mechanism. Instead, he could have been asking, as Kepler did, whether there was some deeper reason why those forces have the form they do. In the *Principia,* Newton investigated other laws of force than the inverse-square law, as if looking over God's shoulder while He chose between alternative laws of nature. Newton proved that some alternative force laws would cause spiral orbits, which perhaps had deterred God from choosing them. With great prescience, Newton noted that a slight shift in the perihelion of Mercury (the point of its closest approach to the Sun) would give crucial evidence that some component of the gravitational force did not vary purely as the inverse square. This was the exact point where Einstein's new considerations entered in, for his revised laws of gravity yielded just such a deviation from the pure inverse square law manifest in Mercury's perihelion. Newton not only set forth an accurate law of universal gravitation; he also anticipated the kind of phenomena that would indicate that his law needed modification.

Newton's commitment to alchemical work strongly suggests that he had great hopes of finding something even greater than the mathematical principles of nature. Besides the million words he wrote on this subject, there is the haunting image that his amanuensis, Humphrey Newton, recorded: Newton worked incessantly during six weeks each spring and fall "in his Elaboratory, the Fire scarcely going out either Night or Day, he sitting up one Night, as I did another, till he had finished his Chymical Experiments, in the Performances of which he was the most accurate, strict, exact: What his Aim might be, I was not able to penetrate into, but his Pains, his Diligence at those set Times, made me think, he aim'd at something beyond the Reach of humane Art & Industry." Although Humphrey is regarded by some modern scholars as a faithful ignoramus doing Newton's incomprehensible bidding, the scene he has captured is unforgettable. Ironically, we are in the same quandary as Humphrey, not really sure what Newton was doing on those fiery nights. As great as was the eros of discovery Newton tasted during the years he was finishing the *Principia,* his alchemical fascination seems no less intense and not less impassioned. Moreover, it overlapped that period of intense mathematical work and even outlasted it.

As it is too facile to read Newton's story through the myth of Oedipus, it is also too easy to read him as Faustus. If anything, he wanted to sell his soul to God, not the devil, to gaze into the incandescent depths of divinity. In his alchemical work there is scarcely any mention of its traditional apex and goal, the Philosopher's Stone and the multiplication of metals; if he pursued the traditional topics of metallic transmutation, it seems he was doing so to check and test the received texts and formulae, but his own goal was "something beyond the Reach of humane Art & Industry." Humphrey's artless words may capture something of it: not gold fever, nor even the quasi-divine hieroglyphics of mathematics, but a direct contact with the living, potent forces of nature. Newton's language in his alchemical writings bears this out, for he uses the vivid traditional images of the art with great freedom. Those images include many erotic symbols, of Venus and Mars ensnared in a net, caught *in flagrante delicto,* of chemical copulations and hierogamies. Except for his charged friendship with Fatio de Duillier, Newton's life was singularly devoid of any erotic involvement; he died a virgin (as he informed his physician on his deathbed, and as Voltaire informed Europe). Thus one cannot help notice his fluent use of these erotic metaphors. Of course, he may have done so purely conventionally, using the traditional locutions as a modern chemist speaks of "mercury" instead of "quicksilver." On the other hand, there are passages in which Newton seems to use the mythical language not conventionally, but to grasp chemical realities. For instance, at one point Newton relies on the myth of Saturn swallowing and regurgitating his children to suggest to him the next steps in a difficult alchemical process.

As guarded and impassive as his style is in the *Principia,* his alchemical writings do not hesitate to evoke the seething, explosive immediacies of his Elaboratory. Describing the arduous preparation of the "star of regulus" (in modern terms, a star-shaped crystalline form of antimony), Newton attests that

> I know whereof I write, for I have in the fire manifold glasses with gold and this mercury [specially prepared by amalgamation]. They grow in these glasses in the form of a tree, and by a continual circulation the trees are dissolved again with the work into new mercury. I have such a vessel in the fire with gold thus dissolved, where the gold was visibly not dissolved by a corrosive into atoms, but extrinsically and intrinsically into a mercury as living and mobile as any mercury found in the world. For it makes gold begin to swell, to be swollen, and to putrefy, and to spring forth in sprouts and branches, changing colors daily, the appearances of which fascinate me every day. I reckon this a great secret in Alchemy . . .

With his own eyes he beholds extraordinary metamorphoses: tree-shaped crystals emerge and dissolve, mercury and gold swell and sprout. He is fascinated precisely because these manifold changes denote something really *alive,* not the dead passivity of matter according to a mechanical description. A prominent strain of alchemy treated nature as living, generated by the activating agency of spirit and the copulations of male and female principles. At times these active principles are described as operating through attraction and are called "magnets," a claim that Newton noted carefully. Newton's language shows that he was deeply moved by the phenomena themselves, and most of all by their quickening.

It is just this sense that informs Newton's image of the ocean, for the boy plays with dead things, though lovely ones: the pebble and the sea shell, both relics of the ceaseless life of the sea. The "great ocean of truth" lies beyond the boy's ken, and Newton shifts his gaze toward the source of life, past the boy's dreamy absorption. The penultimate paragraph of the *Principia* ends with the comment that "it is enough that gravity really exists and acts according to the laws that we have set forth and is sufficient to explain all the motions of the heavenly bodies and of our sea." There, too, it is "our sea" that is the last-named goal of Newton's quest. In his vision, gravity and its laws do not suffice to convey the life of the ocean that goes beyond its passive response to the moon in tides. There remains the fascinating play of the waves and the unending activity of what Leibniz called *vis viva,* "living force" or energy.

This sense of life persists even in the heart of Newton's mathematical work. Newton had referred to the quantities generated by fluxions as *genitum,* literally "born" or "engendered." He indicates by that term, which he coined, quantities "indeterminate and variable, and increasing or decreasing, as if by a continual motion or flux." To the bare notion of pure quantity Newton adds the crucial element of "continual motion or flux," giving rise to ceaseless increase and decrease. The striking resemblance of his description to the motion of the sea is not fortuitous; at one point, Newton considers a "tremulous body," through which he calculates the speed of "pulses," the waves. Newton balances the passivity of matter with the activity of the forces that move it, including the paradoxical inertia by which bodies actively assert their inactivity. His concern is for the living variable, not the lifeless constant.

Even if Newton's grand mathematical project contained a far-reaching attempt to comprehend the living flux of the world, his oceanic image implied a deeper question about what animates living things. The ocean is greater than the pebble. One does not know whether the active principles and copulations that Newton scrutinized so intently in his alchemical studies offered him some access to that deeper truth or whether alchemy finally seemed in his eyes still a diversion, a game with shells. In his ardent hours, the emergent presence of life returned to remind him of its incomparable inwardness, its ultimate claim.

10

Einstein in the Boat

You imagine that I look back on my life's work with calm satisfaction. But from nearby it looks quite different. There is not a single concept of which I am convinced that it will stand firm, and I feel uncertain whether I am in general on the right track.
—Einstein to Solovine, 1949

In the summer of 1931, Einstein went sailing near his vacation home with a visitor. Their conversation "drifted back and forth from profundities about the nature of God, the universe, and man to questions of a lighter and more vivacious nature. . . ." Suddenly Einstein "lifted his head, looked upward at the clear skies, and said: 'We know nothing about it all. All our knowledge is but the knowledge of schoolchildren.' 'Do you think,' I asked, 'that we shall ever probe the secret?' 'Possibly,' he said with a movement of his shoulders, 'we shall know a little more than we do now. But the real nature of things, that we shall never know, never.' "

Einstein never stopped searching for the secret. Despite his breakthrough with gravitation, quantum theory resisted his grasp. As some levels of secrets are revealed, others remain inviolably kept. Einstein struggled with this divided disclosure all his life. As was the case with Kepler and Newton, his wrestling with Proteus was both ardent and arduous. Though he revered Newton, he did not stay on the shore. Knowing how small his boat was, he set out to try the depths.

His quest began with a compass, which showed him the way in many senses. It disclosed "something deeply hidden behind things" in the world and in his own inward depths; it was his first touchstone. Soon he found another: when he was twelve, his uncle gave him a little book about Euclidean geometry. Einstein called it "the holy geometry booklet," with a certain irony, but also with awe. Unlike other children, when the band played and the soldiers marched by, he would turn away in disgust. In the holy booklet, he recognized genuine authority. Though its revelations concerned mathematical objects, Einstein instinctively extended its domain to include "objects of sensory perception, 'which can be seen and touched.'" As with the compass, something hidden lies behind the sensual. Somewhat earlier, Einstein had also been introduced to the basic ideas of algebra. His Uncle Jakob told him that "algebra is a merry science. We go hunting for a little animal whose name we don't know, so we call it *x*. When we bag it, we pounce on it and give it its right name." The hidden realm is populated; there are unnamed beings whom we can seek and "bag." In this supreme game of hide-and-seek, the child gains the mysterious power to compel the hidden animal to disclose its real name. Einstein felt how much pleasure this hunt gave him, and how much he delighted in finding new ways of pursuing the unknown.

The intersection of these stories illuminates Einstein's path to his later discoveries. The compass disclosed a hidden realm behind things; geometry and algebra let him move in that realm. Paradoxically, Einstein had a strong streak of sensuality: his turn of mind was instinctively visual, rather than verbal or abstract, and he described his experience of thought in terms of physical pleasure: an exciting thought made his hair bristle. Although Einstein confessed that he "twice failed rather shamefully" to sustain a lasting relationship with a woman, he was no stranger to eros; yet he was speaking about himself, as well as Spinoza, when he said that "the universe was his only and everlasting love." His closest companions noted his essential apartness, his detachment from ordinary human bonds. He himself remarked that "I am a 'lone traveler' and have never belonged to my country, my home, my friends, or even my immediate family with my whole heart. In the face of all these ties I have never lost a sense of distance and a need for solitude—feelings that increase with the years." His deepest desires turned elsewhere. Fascinated by the hidden realm behind nature, he struggled to find physical laws connecting the visible with the invisible. It was the interface between them that compelled him. His passion was sailing, riding the ever-shifting surface of the deep.

Characteristically, his wrestling with Proteus took the form of thought-experiments, hybrid experiences that transform the physicality of actual experimentation into mental grappling. At age sixteen, he imagined chasing after a beam of light, running alongside it and trying to visualize how it would look if he could match its speed. That would allow him to see what could not be seen, in the ordinary sense. If light were a wave, would it seem to stop waving when he caught up with it? For someone with a great hunger to see, visualizing the invisible would be the ultimate challenge, the greatest thrill. Of course, the difficulty is that the invisible will not submit to anything like ordinary visualization. His "seeing" would have to transcend itself, and that is where mathematics came in, to extend the range of what he could "see." Logic helps when one's legs are not fast enough;

since light never rests, the wave cannot stop, and therefore we can never catch it. The speed of light must be a limit we cannot reach. In this way, he groped toward a hybrid of thinking and feeling that would take him deeper.

Even as he struggled to see the picture, he had to let go of picture-thinking in favor of mathematical symbols. Einstein hesitated to abandon visualization. He wrote simplified accounts that would render his work more accessible, at least at the logical level. As in Newton's case, he relied on geometry to allow him to *see,* not merely deduce through algebraic symbols, however powerful. To illustrate Einstein's theories, Hermann Minkowski devised diagrams that present a space-time of four dimensions, three of space and one of time; to fit on a two-dimensional page, only one spatial axis and the time axis are usually shown. For instance, to a stationary observer, a rapidly moving object appears contracted in length; this can be pictured as a "tilting" of the moving object into the dimension of time. However, our eye cannot really see this way, since the diagrams go beyond three-dimensional space. These diagrams are shadows of four-dimensional images; they are not "pictures" in any normal sense. Einstein wrote wryly that "the non-mathematician is seized by a mysterious shuddering when he hears of 'four-dimensional' things, by a feeling not unlike that awakened by the occult. And yet there is no more common-place statement than that the world in which we live is a four-dimensional space-time continuum." For him it was a "lived" world, but we have to struggle to know what "living it" means.

Despite his love for geometry, Einstein's quest for the Law took him into an invisible realm of symbols. Einstein was aware of this development and noted its roots in his own personal history. There is some irony here, since he understood that the goal of his work was to free himself from what he called "the chains of the 'merely-personal.'" He began as a child with a precocious sense of "the nothingness of the hopes and strivings which chase most men restlessly through life," amplified soon by his sense of "the cruelty of that chase, which in those years was much more carefully covered up by hypocrisy and glittering words than is the case today." He tried traditional Judaism, but dropped it about the time he encountered mathematics, in "a positively fanatic [orgy of] freethinking." His suspicion about every kind of authority increased. Past the trap of the "merely-personal," a miasma of wishes, hopes, and primitive feelings, lies "this huge world, which exists independently of us human beings and which stands before us like a great, eternal riddle, at least partially accessible to our inspection and thinking. The contemplation of this world beckoned like a liberation, and I soon noticed that many a man whom I had learned to esteem and to admire had found inner freedom and security in devoted occupation with it."

In speaking about one of these men, Max Planck, Einstein amplifies his sense of the oppression that he wished to escape. Speaking in 1918, on the occasion of Planck's sixtieth birthday, he imagines an "angel of the Lord" driving out of the "temple of Science" all those who dwell there only out of ambition and thirst for the superior intellectual sport of science, or those whose interest is directed only toward utilitarian purposes. Planck is one of the few who remain, "somewhat odd, uncommunicative, solitary fellows." What has brought them to the temple? "I believe with Schopenhauer that one of the strongest motives that lead men to art and science is escape from everyday life with its painful cruelty and hopeless dreariness, from the fetters of one's own ever-shifting desires." There is a note of desperation, of world-weariness and pain, that sharpens the need to escape (which Kepler also expressed), contrasted against the beauty of the "world of objective perception." Einstein shared with Schopenhauer the view that the perceptible world of appearances and desires is a delusion that can be overcome. There is an undertone of deep sadness also; the strategem may not altogether succeed, and the world may prevail. Nonetheless, in the presence of "our beloved Planck," Einstein continues to speak of "the longing to behold this pre-established harmony" as the source of the endurance of the seeker. His state of mind "is akin to that of the religious worshiper or the lover; the daily effort comes from no deliberate intention or program, but straight from the heart." As Bacon had foreseen, eros turns from the "merely-personal" to the quest for hidden harmonies.

The path that Einstein discerns begins in personal need and leads to a vision beyond the shadow of the "merely-personal." Without this deep personal need, the journey from "the narrow whirlpool of personal experience" to the heights of theoretical vision may seem strange, distant, and abstract. There is also a religious sense in Einstein's account. In some of his accounts, Einstein stressed what he called a "cosmic religious feeling," in which "science not only purifies the religious impulse of the dross of its anthropomorphism but also contributes to a religious spiritualization of our understanding of life." In those accounts, Einstein identifies a ground of religious feeling that transcends the different religious traditions and is consistent with the transcendence of individuality or anthropomorphism in all its forms, particularly through devotion to science. However, there is an interesting resemblance to the rhetoric and imagery of Biblical religion. The "religious paradise of youth" is lost, but a new paradise takes its place, even though "the road to this [new] paradise was not as comfortable and alluring as the road to the religious paradise." Indeed, the "angel of the

Lord" has not swept out of the temple the unworthy initiates with whom Einstein keeps ironic and somewhat uneasy company. The new paradise is not simply absorption in scientific work, but requires also a kind of religious devotion and purity of intent only to be found in a few.

The echo here of Eden is unmistakable, even of the angel with a flaming sword. There is a chosen people, the true seekers of the company of Planck. Most of all, there is the theme of liberation. Einstein speaks of the "chains of the 'merely-personal,'" from which the contemplation of the riddle "beckoned like a liberation." The individual lost in the "narrow whirlpool of personal experience" is enslaved by the state, through its authoritative deceptions and "education-machine." The unpicturable quality of modern physics also reflects the transcendent provenance of its Law. Einstein sought liberation from the "merely-personal" in the form of "picture-thinking." Planck wrote that he "had always regarded the search for the absolute as the loftiest goal of all scientific activity." Looking back over the mathematical physics of Newton and his successors, Planck discerns "an ever extending emancipation from the anthropomorphic elements" through "the elimination of the individuality of the physicist."

The equations of those theories express what is absolute and invariant in any physical situation, leaving aside the "merely-personal." Einstein sought general principles that regulate particular laws. For example, the generalized principle of relativity regulates the character of any possible physical laws by requiring that they "must be of such a nature that they apply to systems of reference in any kind of motion." As a Law of Laws, it eludes visualization when it reaches the greatest heights of generality; there only the gnomic characters of symbolic mathematics are adequate. Did not the Lord name Himself in the form of an *equation,* "*I am what I am*"? His Name takes a form that transcends human names, just as algebra transcends ordinary sentences. Did not the Lord forbid graven images? Mathematical equations mirror the world by transforming it into symbolic structures. As he generalized and deepened his theory, Einstein sought special mathematical objects that could form invariant equations. Studying mathematics, he realized he needed tensors to build his general theory.

Tensors had already been used in mechanics to express stress and strain. For instance, a "stress tensor" records how much a metal beam bearing a load is stressed and twisted. Similarly, a tensor can measure the amount a light ray would bend in traveling in different directions through a crystal. Matter is generally *twisted,* and the tensor records its distortion. One remembers Proteus, held in Menelaos's grip, disclosing his myriad forms. Einstein applied this image to space-time itself, as if it were a kind of crystal that was distorted by the presence of nearby matter. He devised the "metric tensor" to be a Protean summary of all the twists and turns that a beam of light (or of particles) would undergo passing through a given region of space and time. Using the symbolic power of tensors, Einstein wrote an equation expressing the effect of matter on space and time; his expression was completely general, free of the "merely-personal" limitations of a single observer. Einstein was able to hold Proteus even when he assumed the ghostly shape of space and time.

Einstein's equation gives predictions that go beyond Newton's gravitational theory; it seemed that the center of the labyrinth was at hand. Einstein spent his last four decades searching without definitive result for a unified theory of gravitation and electromagnetism that would also subsume the quantum theory, which had been developing rapidly during those years. Although Planck had first introduced the idea of the quantum in 1900, and Einstein himself had radically extended that idea in 1905, the "new quantum mechanics" of 1925–1926 must have been on Einstein's mind in the boat; he wrestled with it for the rest of his life. Even today, quantum theory remains enigmatic. Every attempt to visualize it leads to significant misunderstandings. Paul Dirac noted that "the new theories, if one looks apart from their mathematical setting, are built up from physical concepts which cannot be explained in terms of things previously known to the student, which cannot even be explained adequately in words at all."

Though every theory rests on some premises, quantum theory seems to rest on grounds that are particularly troublesome. Einstein questioned whether the theory really was as complete an account as possible of what man could know about nature. Most of all, he objected to Heisenberg's uncertainty principle, that one could not know both the position and the momentum of a particle with perfect exactitude at the same time. Einstein joked that God does not play dice, meaning that the transcendent Law leaves nothing indeterminate and owes nothing to mere chance. The last half of his life was devoted to a quest to purify that Law from the cloud of indeterminacy. He was extraordinarily patient as he wandered in a wasteland of unfulfilled expectations for those forty years. Most contemporary physicists have not sustained Einstein's reservations about quantum theory. Perhaps Einstein needed Bacon's counsel about questioning one's beloved theories most searchingly, about not being deceived by desire. Yet Einstein achieved his greatest insights through relying on his inner convictions about the character of God's Law. It was very hard for him to accept Niels Bohr's pointed criticism: "Who are *you* to tell the Lord what He may not do?"

Einstein shared his inner conviction with his favorite philosopher, Benedict Spinoza. He even called himself a "disciple" of Spinoza, as if they were alter egos. "Although he lived three hundred years before our time, the spiritual situation with which Spinoza had to cope peculiarly resembles our own. The reason for this is that he was utterly convinced of the causal dependence of all phenomena, at a time when the success accompanying the efforts to achieve a knowledge of the causal relationship of natural phenomena was still quite modest." They both advocated a transformed monotheism uniting Nature with God. Both held that there is no arbitrariness in the world. This rigorous sense of causality goes back in Spinoza to the most fundamental level, to his discussion of the primal nature of "substance," its uniqueness and infinitude, and its identity with God. From there Spinoza in his *Ethics* uses the geometric manner of propositions and proofs to argue that "nothing in the universe is contingent, but all things are conditioned to exist and operate in a particular manner by the necessity of the divine nature." For similar reasons, Einstein adhered to perfect and complete determinism in physics. The new quantum theory "would mean the end of physics" because it abandoned this determinism. "I believe in Spinoza's God who reveals himself in the orderly harmony of what exists, not in a God who concerns himself with fates and actions of human beings." Einstein connects the theme of transcending the "merely-personal" with the consummate harmony of the world; the pursuit of such harmony both requires and grants immunity from the quagmire of "merely-personal" life. Spinoza was a heretical hero, a saint excommunicated by his own people: Einstein valued Spinoza's apartness as he treasured his own. The tension between the abolition of individuality in the face of law and the determinate fate of each separate individual is at the heart of Einstein's struggle with quantum theory. Both are demands that Einstein lays at the foundation of the Law.

Einstein's musings in the boat reflect his conviction that nature is completely determined. To deny or weaken this principle, Einstein held, not only implied a limitation in human knowledge, but reflected on nature and God. Returning to his meditations in the boat, he conceded that we will never know the innermost secret of things. Our faculties are so limited that it is incredible that we can reach as far as we do. The quote at the beginning of this chapter showed Einstein, like Newton, acknowledging the smallness of knowledge and the magnitude of doubt. However, quantum theory was asking Einstein to give up not only *our* limited prospects, but also the far more important notion that such perfection exists somewhere, even if finally inaccessible to us. To put it in Einstein's characteristic way, *God* at least must know perfectly the positions and velocities of each particle, even if *we* never can. Einstein would not give up the inviolate perfection of the Law itself, however inadequate our knowledge. To him that would have been an abandonment of natural philosophy as he understood it. It is as if the labyrinth of nature were to dissolve into a centerless, endlessly shifting mist. Better to go sailing than pursue such phantasms.

Einstein tried for years to concoct some kind of thought experiment that would sharpen these difficulties to the point that others would be persuaded to join him. In a continuing dialogue, Niels Bohr was able to find some crucial flaw or omission in Einstein's examples. Yet Bohr could not deny Einstein's central point: quantum theory involves what Einstein called "spooky action at a distance," bizarre possibilities of knowing the state of one particle based on information gathered from another, distant particle. How could two distant particles be in effective contact with each other? In the succeeding years, the spookiness has not dispersed. Quantum theory predicts correlations even between distantly separated particles; experiment confirms these effects, which do not involve transmission of information faster than the speed of light. Einstein was quite right that quantum theory is weird; nature just *is* that strange. The crux of this strangeness is that quantum particles radically lose their individual identity; they can merge and interfere. This utter indistinguishability or identicality stands in the way of Einstein's determinism, which depends on the individual identity and distinguishability of each particle. Ironically, Einstein missed the implications of this loss of identity, even though he wanted to transcend the "merely-personal." The further irony is that Spinoza's vision of the ultimate merging of identity in the divine substance might have reconciled Einstein to these radical directions of quantum theory. In eluding individual identity, Proteus finally slipped through his grasp.

Full discussion would call for another book; for now, the important point is that quantum theory operates on two levels. There is a deeper dimension here that brings to mind the Egyptian labyrinth, once "a wonder past words" (now long vanished) that surpassed the pyramids, even older and vaster than the Cretan maze. The Egyptian labyrinth had *two* stories, each over a thousand rooms; the underground level was forbidden, because of the tombs of the kings and sacred crocodiles. Likewise, the quantum maze has an outer level, accessible to experiment but indeterminate, as well as an inner level of determinism, inaccessible to direct observation but mathematically knowable. As Max Born pointed out, the mathematical laws of the theory are completely deterministic, but the observable quantities are *not* deterministic: if something is not observable, it can be certain, but if it is observable, it is not certain! The reason is that the act of observation uncontrollably affects the system, at least on the atomic level, and the inner determinism remains hidden. Einstein could not accept such a split between certainty and observability. He held fast to the only clue he had: the Law must be complete and perfect. Though he fully recognized the magnitude of the accomplishments of quantum theory, and how difficult it would be to subsume them in some new, purely deterministic theory, Einstein was not going to be turned away from the center of the labyrinth. Every attempt to continue Einstein's program has so far failed, both his own inconclusive efforts and later attempts to craft deterministic, "hidden variable" theories. The majority opinion is reconciled to quantum theory and uses it to seek a unified theory of the fundamental interactions. The search for the Law that Newton and Einstein expected at the center of the labyrinth has turned more and more to symbolic mathematics, whether in general relativity or the infinite-dimensional "geometry" invoked in quantum theory.

The parallel between cryptanalysis and the decryption of nature continues to be illuminating. As cryptology developed, it used more intricate substitutions and transpositions. Already in the time of Viète, polyalphabetic ciphers began to be used, in which each letter of the plain text is enciphered with a *different* cipher alphabet. Often that repertory of alphabets is governed by a "key," a word or phrase whose letters tell which alphabet to use for each succeeding letter. Such a cipher is far more resistant to cryptanalysis than the simpler monoalphabetic ciphers that Viète solved. However, as Viète's "infallible rule" suggests, even a polyalphabetic cipher can be solved by general procedures, later set forth by Friedrich Kasiski in 1863. These methods might make one think that any cipher is finally soluble, but that is not so. It was only in 1917 that a truly unbreakable cipher was invented: the "one-time pad." This is a keyed cipher that uses a *random* key (for instance, a table of random numbers, which the sender and receiver share). Most important, the list of random numbers used as the key must be as long as the message to be sent and must be used *only once.* Any decryption requires accumulating enough enciphered text to deduce the key, but if the key is nonrepeating and random, there is no way to deduce it. Despite its impregnability, the one-time pad was not much used until recent years, mainly because of the awkwardness of the key required; it was used by such spies as Rudolf Abel, who had tiny one-time pads of random numbers that he would use and then destroy. Other critical transmissions, such as the Washington-Moscow "hot line," also rely on it.

Meanwhile, cryptanalysis continued to unite with algebra. Complex rotor enciphering machines were invented in 1917, in the United States, and independently within the next few years, in the Netherlands, Sweden, and Germany; their operation can be symbolized by algebraic matrix equations. These machines developed into the famous German "Enigma" and Japanese "Purple," which played such a fateful role in World War II. Despite the enormous complexity of these ciphers, their equations were soluble, a fact of immense consequence to the outcome of the war. The algebraic approach can break these polyalphabetic ciphers, but fails against the one-time system.

The crucial difference is that the algebraic equations of the one-time system are *not soluble;* there are vastly more unknowns than equations, since there is no way to deduce the endless, unrepeating stream of random numbers in the key. It is this brute algebraic fact that seals its unbreakability. In contrast, the "Enigma" and "Magic" decryptions are equivalent to solving for unknowns fewer in number than the given set of equations; the cipher machines were limited in their possibilities and did not rely on randomness.

Quantum physics discloses an endless source of random numbers; radioactive nuclei or Josephson junctions can give a random series of *0*s and *1*s ideally suited to a computer. Their randomness is confirmed by the failure of extensive experimental attempts to find some pattern in these phenomena. This is just the dice-throwing that Einstein found so vexing, and it provides an ideal way to implement such unbreakable one-time ciphers. A computer can take random digits from the quantum source and use them to encipher the message. The emergent techniques of the new forms of "quantum cryptography" harness quantum dice-shaking to the one-time system. *Quantum theory, in itself, is a one-time system,* yoking uncontrollable external randomness to internal determinism. What this means is that *nature's cipher is unbreakable, though intelligible mathematical law operates underneath that cipher.* Nature's secret is encrypted in randomness so complete that she herself could not solve it. After all, the crux of the one-time system is that the key must be perfectly random and endless, and hence completely unknowable, *even to its senders,* who may possess it as a list but who do not—cannot—understand its order, because it has none. Consider what keys are knowable. Finite keys are clearly knowable—for example, a motto or even nonsense phrases such as were used in the first polyalphabetic ciphers. But any finite key is breakable, since it will reduce to a limited, soluble set of equations. Even an *infinite* key is breakable if it repeats in finite units or has any finite order or pattern. If it is knowable, it is breakable, even though endless. Thus God would have to "know" an infinitude of random digits to possess the key, and such randomness is just what so exercised Einstein. Without going as far as Einstein did in defense of eternal Law, it is evident that such randomness is scarcely compatible with *any* sort of mind, much less a divine one. The only way of getting truly random numbers is total stupidity, for any scintilla of mind will pollute the randomness with some taint of order.

No wonder Einstein could not endure such a thought. It attacks the fundamental ordering of the world. Yet though randomness inheres in outward physical appearances, the inner level of the theory remains perfectly determined and lawful. Observable properties remain the fertile domain for the investigation and interpretation of nature; the hidden interiority of things remains elusive. Aristotle long ago pointed out the irony in "explaining" something familiar in terms of something obscure. What remains is the challenge to *connect* the disparate realms of our experience. Can we relate the boiling point of water to other measurable quantities? Is there some connection between the smallest observable constituents of matter? These may be answerable questions.

The hiddenness of nature insures that its depths will never be exhausted. Both Einstein and Newton felt they were children in the face of such immensity. With Kepler, they felt that even a small step in understanding represented contact with something beyond human power. They wrestled with desire and delusion as they struggled to break the code. Finally, their intense struggle to read the strange message evoked a profound internal reaction, which Einstein already sensed in his first encounter with the compass. As a child, Einstein trembled and grew cold because he felt an inward stirring that answered the outward pointing of the needle. He felt not only the enigmatic wonders of nature, but also a new depth that answered within him. The human significance of science lies in the quest for those depths. The search for "something deeply hidden behind things" not only promises strange and beautiful insights; the attempt to reach past the merely-human evokes a new depth in human nature.

Kepler glimpsed an endless vista in the sky; Newton sensed infinite depth in the ocean of truth; Einstein recoiled from a randomness that eludes determination. The image of a maze guarding a single, isolated center may no longer be adequate. The secret is not in one spot, but in the whole design, the hidden signature of the infinite, whose center is nowhere, and everywhere.

Notes

Prelude

"Something deeply hidden behind things": See Einstein's "Autobiographical Notes" in Paul Arthur Schilpp, *Albert Einstein, Philosopher-Scientist* (New York: Harper, 1959), p. 9. The detail about his trembling and growing cold is related in the biography written by Einstein's son-in-law Rudolf Kayser, under the pseudonym Anton Reiser, *Albert Einstein, a Biographical Portrait* (New York: A. and C. Boni, 1930), p. 25.

The "sedimentation" of meaning: This influential term was used by Edmund Husserl in *The Crisis of European Sciences and Transcendental Phenomenology* (Evanston: Northwestern University Press, 1970), p. 52. See also Jacob Klein, *Lectures and Essays* (Annapolis: St. John's College Press, 1985), pp. 77–84, who notes that, though mostly forgotten, the sediment has not completely disappeared from our understanding and can be reactivated.

The ancient view of the hiddenness of nature: Martin Heidegger also makes this connection, in a rather different way; see his *Early Greek Thinking* (New York: Harper and Row, 1975), p. 113 ff. However, Heidegger, in his aversion to modern science, does not really address the kind of connections I will explore.

"Nature loves to hide": Heraclitus of Ephesus was an early Greek thinker, who lived about 500 B.C. This is Diels-Kranz Fragment B17, in *The Presocratics,* ed. Philip Wheelwright (New York: Odyssey Press, 1966), p. 70. "Unless you expect the unexpected": Heraclitus Fragment B19, *Presocratics,* p. 70.

"Truly, thou art a God who hidest thyself": Isaiah 45:15. Blaise Pascal emphasized the hiddenness of God, but in the context of radically fallen nature; see Leszek Kolakowski, *God Owes Us Nothing* (Chicago: University of Chicago Press, 1995), pp. 113–197 (at pp. 148–149), Lucien Goldmann, *The Hidden God* (London: Routledge, 1964), pp. 220–259, and Ioan P. Couliano, *Eros and Magic in the Renaissance* (Chicago: University of Chicago Press, 1987), pp. 207–208. On the earlier religious context of science, see Amos Funkenstein, *Theology and the Scientific Imagination* (Princeton: Princeton University Press, 1986) and Roger French and Andrew Cunningham, *Before Science: The Invention of the Friars' Natural Philosophy* (Aldershot: Scolar Press, 1996), pp. 70–109, 269–274. For developments in the seventeenth century, see Jean-Luc Marion, "The Idea of God" in *The Cambridge History of Seventeenth-Century Philosophy,* ed. Daniel Garber and Michael Ayers (Cambridge: Cambridge University Press, 1998), vol. 1, pp. 265–304.

The hermetic tradition: See Brian P. Copenhaver, *Hermetica* (Cambridge: Cambridge University Press, 1992) and Frances A. Yates, *Giordano Bruno and the Hermetic Tradition* (Chicago: University of Chicago Press, 1964). Yates's thesis that modern science was born out of magic and hermetic thought is fundamentally mistaken, I think, since it does not distinguish the character of magic (which is fundamentally traditional and esoteric) from that of science (which breaks away from tradition and tends towards disclosure).

Alchemy and the "secrets of nature" in early modern Europe: See the excellent treatments by Lorraine Daston and Katharine Park, *Wonders and the Order of Nature* (Cambridge: Zone Books, 1998), Pamela Smith, *The Business of Alchemy* (Princeton: Princeton University Press, 1994), and William Eamon, *Science and the Secrets of Nature* (Princeton: Princeton University Press, 1994).

Changing views of Bacon: For a brilliant rebuttal of Macaulay's charges, see Nieves Mathews, *Francis Bacon: The Story of a Character Assassination* (New Haven: Yale University Press, 1996); see also Antonio Pérez-Ramos, "Bacon's legacy" in *The Cambridge Companion to Bacon,* ed. Markku Peltonen (Cambridge: Cambridge University Press, 1996), pp. 311–334.

The "Age of Bacon": This is the appellation of Gernot Böhme, *Am Ende des Baconischen Zeitalters* (Frankfurt-am-Main: Suhrkamp, 1993).

Themes in scientific thought: For a trail-blazing collection of thematic studies, see Gerald Holton, *Thematic Origins of Scientific Thought* (Cambridge: Harvard University Press, 1988) [second edition].

Insider vs. outsider history: See Steven Shapin and Simon Schaffer, *Leviathan and the Air-Pump* (Princeton: Princeton University Press, 1985), pp. 4–6. Social approaches to the understanding of science often tend to neglect the realities of nature in favor of social or political explanations; in so doing, they lose the center of concern for the scientist.

Chapter 1: The Hard Masters

William Gilbert, *De Magnete* (New York: Dover, 1991); citations of this work will refer to the page number of this edition. I have silently used capitalization to distinguish between the Earth and the Aristotelian element, earth; here and elsewhere I have modernized spelling. Duane H. D. Roller, *The De Magnete of William Gilbert* (Amsterdam: Menno Hertzberger, 1959) surveys Gilbert's historical background.

Chinese knowledge of magnetism: See Joseph Needham, *Science and Civilization in China* (Cambridge: Cambridge University Press, 1962), vol. IV, part 1, pp. 229–334. The Chinese already had a sense of organic relation between the lodestone and the Earth. For helpful selections from Greek and Roman writers, see *A Source Book in Greek Science*, ed. M. R. Cohen and I. E. Drabkin (Cambridge: Harvard University Press, 1958), pp. 310–315. For a general history of magnetism, see Gerrit Verschuur, *Hidden Attraction* (New York: Oxford University Press, 1996).

Gilbert and the terrella: He seems to have gotten the idea for this from Peter Peregrinus (1269), who shaped a lodestone into a globe, although he did not use the term terrella. Peregrinus noted "that this stone bears in itself the likeness of the heavens," and considered magnetism to have a celestial, rather than terrestrial, source. There is a very useful general selection of writings (including Peregrinus) in *A Source Book in Medieval Science*, ed. Edward Grant (Cambridge: Harvard University Press, 1974), pp. 367–376. Another important source (though Gilbert says that it is "full of most erroneous experiments," 11) is Giovanni Battista della Porta, *Magia Naturalis* (1589), available as *Natural Magick*, ed. Derek J. de Solla Price (New York: Basic

Books, 1959), pp. 190–216. For Gilbert's relation to his sources, see Edgar Zilsel, "The Origins of William Gilbert's Scientific Method," *Journal of the History of Ideas* 2, 1–32 (1941) and Peter Dear, *Discipline and Experience* (Chicago: University of Chicago Press, 1995), pp. 159–161.

Magnetic variation: Here Gilbert drew material from the mariners Robert Norman, *The newe Attractiue* (1581) and William Borough, *A Discovrse of the Variation of the Compasse, or Magneticall Needle* (1596), included in *Rara Magnetica,* ed. G. Hellman (Nendeln/Lichtenstein: Kraus Reprint, 1969), pp. 83-105, 107–115. In accord with his preference for love over war, Gilbert seems to assimilate others' work without anticipating disputes about priority or indebtedness.

Gilbert and Copernican astronomy: See Gad Freudenthal, "Theory of Matter and Cosmology in William Gilbert's *De Magnete,*" *Isis* 74, 22–37 (1983).

Gilbert's fidelity to iron: An ironic rebuke to the "vain imagination of chemists to deem that nature's purpose is to change all metals into gold" (42). The starry diamond only glistens like the stars; the magnet guides men more surely than the stars (217–218).

Contemporary metallurgy: Gilbert quotes from the most important contemporary source, Georgius Agricola, *De Re Metallica* (1556), available in the translation by H. C. Hoover and L. Hoover (New York: Dover, 1912). See Pamela O. Long, "The Openness of Knowledge: An Ideal and Its Context in 16th-Century Writings on Mining and Metallurgy," *Technology and Culture* 32, 318–355 (1991).

The metal speaking from the retort: This papyrus is attributed to Zosimos of Panopolis (Akhmim) in Egypt (ca. A.D. 300); see E. J. Holmyard, *Alchemy* (New York: Dover, 1990), p. 29, and Mircea Eliade, *The Forge and the Crucible* (New York: Harper & Row, 1962), pp. 149–152. C. G. Jung, *Alchemical Studies* (Princeton: Princeton University Press, 1970), pp. 57–108, 328–333 includes a translation of the Zosimos papyrus and discusses the alchemist's purificatory ordeal.

The Maker's perspective on knowledge: See Antonio Pérez-Ramos, *Francis Bacon's Idea of Science and the Maker's Knowledge Tradition* (Oxford: Clarendon Press, 1988), pp. 48–62.

Gilbert's medical career: John Butterfield, "Dr Gilbert's magnetism," *The Lancet* 338, 1576–1579 (1991).

Form and magnetism: Gilbert does not speculate how that "form" is constituted; he does not attempt any explanation in terms of atoms or microscopic domains within the magnet. He seems to reject discrete atoms in favor of a "continuous and homogeneous" substrate.

Ancient imagery of eros and magnets: Pliny, *Natural History* 26.126. Lucretius invokes erotic desire as a governing image for magnetic attraction, recalling his opening invocation of Venus. He calls the atoms seeds (*semina*) and, like Pliny, speaks of the lodestone as male and the receptive iron, female (*On the nature of things* 6.1002–1008).

Gilbert's critique of "attraction": He may also be thinking of the phenomenon of magnetic *repulsion* (28–31), which obviously would not accord with a primal notion of attraction.

Metallurgy and violence: Pliny considered iron "the most deadly fruit of human ingenuity. For to bring death to men more quickly we have given wings to iron and taught it to fly." (*Natural History* 34.39) See Carolyn Merchant, *The Death of Nature* (San Francisco: Harper & Row, 1980), pp. 29–41. Agricola identifies iron with the fire Prometheus gave to men, against the gods' command. Though acknowledging legal uses of torture, Agricola defends metallurgy from the charge of being a purveyor of torture implements (p. 17).

Citations from Gilbert in text: 3 (light of experience); l (hidden things that have no name); xlix (new style of philosophizing); 66 (potency of Earth's core); 179 (Earth, the mother); 179 (a deadweight planted in the center of the universe); 108 (effect of fire); 34 (true child of Earth; hard masters); 184 (digging lodestone out of vein; inner parts of Earth); 108–109 (not iron, but something lying outside); 108 (foul bed or matrix); 98 (bride and spouse); 113 (rush towards union); 35 (uterus); 59 (twins); 36 (blood and semen); 58 (fatal effects); 101 (feeding on iron filings); 214 (rebirth of magnet); 97–98 (tyrannical violence); 321 (hateful tyranny); 95 (electric coition); 119 (mighty power); 147 (mutual love); 109 (not indeterminate and confused).

Chapter 2: Wrestling with Proteus

All citations from Bacon refer to volume and page number in *The Works of Francis Bacon,* ed. James Spedding (London: Longmans and Co., 1857–1874; reprint: New York: Garrett Press, 1968); the first seven volumes include the *Works,* and the succeeding volumes comprise *The Letters and Life of Francis Bacon.* A new edition is in progress from Oxford University Press; a valuable collection of the English works is available in *Francis Bacon,* ed. Brian Vickers (Oxford: Oxford University Press, 1996), which includes a fine introduction (pp. xv–xliv). See especially John C. Briggs, *Francis Bacon and the Rhetoric of Nature* (Cambridge: Harvard University Press, 1989); I am indebted to this outstanding work on many points. See also Perez Zagorin, *Francis Bacon* (Princeton: Princeton University Press, 1998).

For complete references for this chapter (especially about jurisprudence and Bacon's scientific vision), see my essays "Wrestling with Proteus: Francis Bacon and the 'Torture' of Nature," *Isis* 90:1, 81–94 (1999) and "Nature on the Rack: Leibniz' attitude towards Judicial Torture and the 'Torture' of Nature," *Studia Leibnitiana* 29, 189–197 (1998).

"Vexation": The phrase "vexation of mind [or spirit]" is common; a "vexed question" requires prolonged examination. In contrast with the physical torments of the rack, vexation acts in the inwardness of the soul. Bacon would not allow "double vexation" in his court of Chancery, meaning unjustified legal actions or harassment (7.762).

"Nature falls silent on the rack": See Johann Wolfgang von Goethe, *Scientific Studies,* tr. Douglas Miller (New York: Suhrkamp, 1988), p. 307.

Bacon and torture: Mathews clarifies his reluctant participation and the reservations he expressed to the king in *Francis Bacon,* pp. 283–294. For Coke's attitude, see pp. 284–285. For King James and witchcraft trials, see his *Daemonologie (1597)* (New York: Barnes and Noble, 1966).

On Bacon's religious views: See John Briggs's essay on "Bacon's Science and Religion" in *Cambridge Companion to Bacon,* pp. 172–199. Despite the claims of later writers that he was not a Christian, Bacon repeatedly asserted that he was one; he composed a profession of Christian faith (7.215–226), sacred meditations (7.243–254), and several prayers (7.259–262).

Bacon's use of classical myth: See Charles W. Lemmi, *The Classic Deities in Bacon* (New York: Octagon Books, 1971), pp. 91–98 (on Proteus). Lisa Jardine, *Francis Bacon* (Cambridge: Cambridge University Press, 1974) treats Bacon's use of parable on pp. 179–193. See Briggs's insightful readings in *Francis Bacon,* pp. 32–40 (Proteus), 19 (Prometheus).

The bonds of Proteus: The "bonds" are already in Virgil's *Georgics* (4.399–405), though absent in Homer even in its 1537 Latin translation. Briggs notes that in all these accounts, including Bacon's, the interrogators "overpower [Proteus] without violating his divinity." (*Francis Bacon,* p. 35.) See also the helpful discussion in Eamon, *Science and the Secrets of Nature,* pp. 283–291.

Self-fashioning: See, for instance, Stephen Greenblatt, *Renaissance Self-Fashioning* (Chicago: University of Chicago Press, 1983). Those who advocate self-fashioning as a master key of interpretation do not seem to recognize Bacon's insistence that the struggle with Proteus goes far beyond what the self alone can fashion; the god's disclosures exceed the seeker's anticipations.

Bacon and the alchemists: Bacon thought that alchemists often overheat their work and recommends temperate heating; he says that natural gold is formed "where little heat cometh" (2.449). See Stanton J. Linden, "Francis Bacon and Alchemy: The Reformation of Vulcan," *Journal of the History of Ideas* 35, 547–560 (at 558) (1974).

Citations in the text (from Bacon unless otherwise noted): 4.29 (epigram; vexations of art); Gilbert, p. 60 ("poisoning" lodestones); 5.403–405 (Bacon on the lodestone); 4.161–162 (genuine forms); 4.263 (interrogatories); 11.280 (the King's man); 10.114 (torture used for discovery); 7.243–246 (exaltation of charity); 7.245 (the deeps of Satan); Bacon's *Masculine Birth of Time,* in Benjamin Farrington, *The Philosophy of Francis Bacon* (Liverpool: Liverpool University Press, 1964), p. 62 (nature as wife); *Odyssey* 4.365–570, Bacon 6.725–726 (Proteus); 4.257 (handcuffs of Proteus); 4.47 (nature working within); 4.202 (too inhuman); 4.199–200 (bodies tormented by fire); 4.324 (deceive nature sooner than force her); 4.32 (nature to be conquered must be obeyed).

Chapter 3: The Wounded Seeker

For further references for chapters 3 to 5, see my essay "Desire, Science, and Polity: Francis Bacon's Account of Eros," *Interpretation* 26: 3, 333–352 (1999). My reading of Oedipus is inspired by and indebted to Briggs, *Francis Bacon,* pp. 13–14, 174, 214.

Bacon and monsters: See the excellent discussion in Daston and Park, *Wonders and the Order of Nature,* pp. 220–231, 260, 290–301; pp. 173–214 treats the emergent significance of monsters.

Bacon and Aristotelian science: See Michel Malherbe, "Bacon's method of science" and Antonio Pérez-Ramos, "Bacon's forms and maker's knowledge," pp. 75–98, 99–120 in *The Cambridge Companion to Bacon.*

Passing "from the Muses to Sphinx": In this distinction between pure research and technology Bacon answers the claim that science is essentially identical with technology; see Martin Heidegger, *The Question Concerning Technology,* tr. William Lovitt (New York: Harper & Row, 1977), pp. 3–36. Roger Shattuck also tends to collapse this distinction in his treatment of the Sphinx in *Forbidden Knowledge* (New York: St. Martin's Press, 1996), pp. 179–180, 323–324.

The ancient myth of Oedipus: See Lowell Edmunds, *Oedipus: The Ancient Legend and Its Later Analogues* (Baltimore: Johns Hopkins University Press, 1985), pp. 9–12. See Sophocles's *Oedipus the King* lines 1033, 809, 1270, and 1525. In Sophocles, the double mark of the pierced feet recurs in the wound Laius gives with a double goad, and in the wound Oedipus gives his two eyes; see Alister Cameron, *The Identity of Oedipus the King* (New York: New York University Press, 1968). Bacon probably did not know Sophocles's text.

Childhood infirmities of scientists: See the work of the sociologist Anne Roe, *The Making of a Scientist* (New York: Dodd, Mead, & Co., 1952), as well as other studies summarized by Gerald Holton in "On the Psychology of Scientists, and Their Social Concerns" in *The Scientific Imagination: Case Studies* (New York: Cambridge University Press, 1978), pp. 229–252.

King James's lameness: See David Harris Willson, *King James VI and I* (New York: Oxford University Press, 1967), p. 16. Bacon himself suffered from an affliction of the heel, perhaps gout; see Lisa Jardine and Alan Stewart, *Hostage to Fortune: The Troubled Life of Francis Bacon* (New York: Hill and Wang, 1999), p. 398.

On Anthony Bacon and Francis: Briggs notes that "it is the maimed hero Oedipus whose life best presages Bacon's history"; see his biographical essay on Bacon, to appear in the new edition of the *Dictionary of National Biography* (for the "impotent feet" see present edition, 1.799b).

On gnawing thoughts: In the essay "Of Friendship," Bacon remarks that those who lack friends "are cannibals of their own hearts" and advises those who are burdened with weighty secrets to remember the "dark, but true" parable of Pythagoras "*Cor ne edito,* 'Eat not the heart'" (6.440).

Citations in the text (from Bacon unless otherwise noted): 6.757–758 (epigram: Sphinx); 4.60 (every student of nature); Proverbs 25:2 (The glory of God); 7.245 (tempting God); 3.394–395 (enchanted glass); Sphinx 6.755–758; "Of Deformity" 6.480–481; Prometheus 6.745–753; 6.469 (force maketh Nature more violent).

Chapter 4: The Creatures of Prometheus

The myths of Vulcan: See Timothy Gantz, *Early Greek Myth* (Baltimore: Johns Hopkins University Press, 1993), pp. 74–78. Bacon kept in his country residence a depiction of Mars caught in a net by Vulcan; see Jardine and Stewart, *Hostage to Fortune,* p. 314.

Bacon's unpublished works: Quotations are from the translation in Farrington, *The Philosophy of Francis Bacon,* pp. 83, 109, 85, 72.

James I and Solomon: Briggs notes that "The extended analogy between James and Solomon would have been a deeply probing model for a scholarly king known for indulgence in pleasure, for his pious reputation, and his willingness to insist upon the divine right of kings." (*Francis Bacon,* p. 40.)

Issues of gender: See Merchant, *The Death of Nature*, pp. 164–191, Evelyn Fox Keller, *Reflections on Gender and Science* (New Haven: Yale University Press, 1985), pp. 33–42, and David F. Noble, *A World Without Women* (New York: Knopf, 1993). For critiques of such readings, see Alan Soble, "In Defense of Bacon," in *A House Built on Sand,* ed. Noretta Koertge (New York: Oxford Unviersity Press, 1998), pp. 195–215 and Iddo Landau, "Feminist Criticism of Metaphors in Bacon's Philosophy of Science," *Philosophy* 73, 47–61 (1998), and my "Wrestling with Proteus." Regarding Bacon's emotional life, see Mathews, *Francis Bacon,* pp. 304–319; Jardine and Stewart, *Hostage to Fortune,* pp. 163, 437, 464–466.

Orpheus and the character of science: My readings are indebted to Briggs, *Francis Bacon,* pp. 1–2, 21, 134–136, 146, 176.

Citations in text (from Bacon unless otherwise noted): C. Kerényi, *Prometheus* (New York: Pantheon, 1963), p. 8 (Goethe epigram); *Iliad* 1.599, 1.591–593, 18.396–397 (Hephaistos; see also Hesiod,

Theogony 927–929 and Pindar, Olympian 7.35–37); *Iliad* 18.418–420 (golden attendants); *Iliad* 18.376–377 (moving tripods); 6.745 (Prometheus making man); 6.736 (Ericthonius); 6.752–753 (Prometheus's assault on Minerva); 6.743–744 (Atalanta); 5.463 (Cupid); 6.740–743 (Dionysus); 6.751 (school of Prometheus); 3.262, 6.764 (Solomon); 6.762–764 (Sirens); 6.720–722 (Orpheus).

Chapter 5: The New Eros and the New Atlantis

Bacon's *New Atlantis:* For a general introduction, see Vickers, *Francis Bacon,* pp. 785–790, Briggs, *Francis Bacon,* pp. 169–174, and "Bacon's Science and Religion," in *Cambridge Companion to Bacon,* pp. 192–197; for other references, see my "Science, Desire, and Polity."

Utopias and eros: The preoccupation with eugenics is a commonplace of utopian works; see Plato's *Laws* 772a and Tommaso Campanella's *City of the Sun* [ca. 1602], included with Thomas More's *Utopia* in Henry Morley, *Ideal Commonwealths* (Port Washington, NY: Kennikat Press, 1968).

The conversion of Bensalem: See David Renaker, "A Miracle of Engineering: the Conversion of Bensalem in Francis Bacon's *New Atlantis,*" *Studies in Philology* 87, 181–193 (1990).

Christian humanism: See Douglas Bush, *English Literature in the Earlier Seventeenth Century, 1600–1660* (New York: Oxford University Press, 1962) [second edition], pp. 35–38 and Herschel Baker, *The Wars of Truth* (Cambridge: Harvard University Press, 1952).

The hidden position of the mother: John Archer notes that "the mother, though excluded from the open display of power, is in the same position in relation to her family that Bensalem is to the rest of the world"; *Sovereignty and Intelligence* (Stanford: Stanford University Press, 1993), p. 148. See also Sharon Achinstein, "How to Be a Progressive without Looking Like One: History and Knowledge in Bacon's *New Atlantis,*" *Clio,* 17:3, 249–264 (at n. 10) (1988).

Population: W. M. S. Russell noted that Bacon anticipated Malthus's connection of population and resources; one infers that he would have been all the more aware that Bensalem should not have suffered underpopulation. See "The Origins of Social Biology," *Biology and Human Affairs* 41, 109–137 (1976). Russell notes that the word *population* in this sense is used for the first time in English in Bacon's essay "Of the True Greatnesse of Kingdomes and Estates" (1612; 6.447).

Freud: See Sigmund Freud, *Civilization and Its Discontents,* tr. James Strachey (New York: 1962).

Citations in text (all from Bacon, *New Atlantis,* 3.127–166, except the following): 4.60 (every student of nature); *Thoughts and Conclusions* in Farrington, *Philosophy of Francis Bacon,* p. 73 (alchemists); 3.167 (versions of bodies); 6.735 (instruments of lust); 6.754 (Icarus); 6.757 (Sphinx); 6.479 (beauty and strangeness).

Chapter 6: The Clue to the Labyrinth

For further references, see my essay "The Clue to the Labyrinth: Francis Bacon and the Decryption of Nature," *Cryptologia* 24 (2000); for Bacon's treatment of encryption, I am indebted to Briggs, *Francis Bacon,* pp. 13–40.

Biblical cryptology: Regarding these methods of "atbash," "albam," and "atbah," see the classic general history of cryptology, David Kahn, *The Codebreakers* (New York: Simon & Schuster, 1996), pp. 78–80. The kabbalah indicated ways in which Hebrew letters could be considered formative elements of creation and also influenced the development of cryptology (pp. 91–92).

Ancient and Islamic history of cryptography: see Kahn, *Codebreakers,* pp. 71–99. Plutarch notes that "it is thought that [Caesar] was the first who contrived means for communicating with his friends by cipher . . ." (*Parallel Lives: Caesar*).

Emblems: Ernst Robert Curtius, *European Literature and the Latin Middle Ages,* tr. Willard R. Trask (New York: Pantheon, 1953), pp. 319–326, 345–347; Michel Foucault, *The Order of Things* (New York: Random House, 1970), pp. 25–44; William B. Ashworth, Jr., "Natural History and the Emblematic World-View," in *Reappraisals of the Scientific Revolution,* ed. David C. Lindberg and Robert S. Westman (Cambridge: Cambridge University Press, 1990), pp. 303–332.

Development of cryptology and diplomacy in early modern Europe: See Kahn, *Codebreakers,* pp. 106–156 (through Antoine Rossignol)

Blaise de Vigenère, *Traicté des Chiffres* (Paris: Abel l'Angelier, 1586), ff. 53r–54v; see also Kahn, *Codebreakers,* pp. 145–148. I have discussed Vigenère's arguments in detail in "Secrets, Symbols, and Systems: Parallels between Cryptanalysis and Algebra, 1580–1700," *Isis* 88, 674–692 (1997). Paracelsus also held that nature is "a vast complex of signs and a ciphered discourse."

Bacon and cryptology: See Archer, *Sovereignty and Intelligence,* pp. 121–131, who notes that "Francis handled the correspondence while Anthony acted as chief decoder" for Essex's ring of spies (p. 126). See also Jardine and Stewart, *Hostage to Fortune,* pp. 55–56, 148, 152, 158, 466.

James I and secrecy: See Jonathan Goldberg, *James I and the Politics of Literature* (Baltimore: Johns Hopkins University Press, 1983), pp. 55–112. Bacon wrote a *History of the Reign of King Henry VII,* whom he described as a great and good king, yet "infinitely suspicious . . . sad, serious, and full of thoughts and secret observations" (6.242–243); see Archer, *Sovereignty and Intelligence,* pp. 133–139.

Nature not encrypted in a human language: See Paolo Rossi, "Hermeticism, Rationality, and the Scientific Revolution" in *Reason, Experiment, and Mysticism,* ed. M. L. Righini Bonelli and William R. Shea (New York: Science History Publications, 1975), pp. 247–273 at 258–259; Briggs, *Francis Bacon,* pp. 13–40; James J. Bono, *The Word of God and the Languages of Man* (Madison: University of Wisconsin Press, 1995), vol. 1, pp. 199–246.

Bacon's "alphabet of nature": See Bacon 4.439, 5.132–133. Bacon's *Abecedarium Naturae* (5.208–211) gives lists of Greek letters, which express in symbolic form "a kind of dictionary, a systematic account, an exhaustive catalogue of the 'letters', primary 'natures' and motions, knowledge of which would enable one to read the language of nature," as Graham Rees notes in "Bacon's Philosophy: Some New Sources with Special Reference to the *Abecedarium novum naturae,*" in *Francis Bacon,* ed. Marta Fattori (Rome: Edizioni dell'Ateneo, 1984), pp. 223–244.

John Dee: See *The Mathematicall Praeface to the Elements of Geometrie of Euclid of Megara (1570)* (New York: Science History Publications, 1975) and Kenneth J. Knoespel, "The Narrative Matter of Mathematics," *Philological Quarterly* 66, 27–46 (1987).

Citations in the text (from Bacon unless otherwise noted): 4.18 (epigram: labyrinth); 3.221 (two books); 8.63 ("Great Cause"); 4.439 (real characters); 4.498 (hieroglyphics); 4.444 (rules for secret writing); 4.445–446 (biliteral cipher); 4.440, Herodotus, *History,* 5.92

(the two tyrants); 4.447 (futile ciphers); 6.689 (parable as ark); 6.695 (veil of fables); 6.756 (Sphinx); 6.698 (not as a device for shadowing); 4.115 (interpretation); 3.503, *Clue to the Labyrinth;* 4.127 (induction); 5.210 (alphabet of nature); 4.40 (as if by machinery); 4.52 (harsh and out of tune).

Chapter 7: To Leave No Problem Unsolved

For further details and references, see my "Secrets, Symbols, and Systems." For translations of the original Viète documents, see my paper "François Viète, Father of Modern Cryptanalysis—Two New Manuscripts," *Cryptologia* 21:1, 1–29 (1997). For general background, see also Kahn, *Codebreakers,* pp. 116–188 (Viète), 157–188 (the era of the Black Chambers).

Bacon's attitude towards mathematics: 3.406, 1.663, 4.448–9 (mathematics as torture); Graham Rees has argued that "the scope of Bacon's mathematical concerns was far wider than is usually granted"; see "Mathematics and Francis Bacon's Natural Philosophy," *Revue internationale de philosophie* 40, 399–426 (1986). Bacon could have known of Viète's work through his friend Nathaniel Torporley, who had been Viète's amanuensis.

Al-Khwārizmi: His work (called in Arabic *Al-jabr wa'l muqabālah*) is readily available in *A Source Book in Mathematics, 1200–1800,* ed. D. J. Struik (Cambridge, MA: Harvard University Press, 1969), pp. 55–60. To give an example from his text in modern notation, if one considers $x^2 + 10x = 39$, by *algebra (al-jabr)* one could add 25 to both sides to "complete the square," yielding $x^2 + 10x + 25 = 64$. This is the same as $(x + 5)^2 = 8^2$ and leads, by taking the square root of both sides, to $x + 5 = 8$, so that $x = 3$. Al-Khwārizmi takes the equation $50 + x^2 = 29 + 10x$ and reduces it by *almucabala,* subtracting 29 from both sides, yielding $21 + x^2 = 10x$, which Al-Khwārizmi writes in the rhetorical form, "There remains twenty-one and a square, equal to ten things."

Islamic alchemy: See E. J. Holmyard, *Alchemy* (New York: Dover, 1990), pp. 60–104.

John Dee: See Peter French, *John Dee: The World of an Elizabethan Magus* (New York: Ark, 1987), p. 5 ff and Nicholas H. Clulee, *John Dee's Natural Philosophy* (London: Routledge, 1988), pp. 122–123,

170–176 (on Dee's magical interpretation of "experimental science"). Though Dee did not accomplish any significant original mathematical work, in 1575, Richard Forster claimed that mathematics was carried and reborn in England on Dee's "Atlas-like shoulders." See also Dee's *Mathematicall Praeface.*

Mathematics and magic: In the thirteenth century, Roger Bacon had described mathematics as "the second division of magic"; he had meant to praise mathematics as well as magic, but from this connection others drew quite different conclusions. See *The Opus Majus of Roger Bacon,* tr. Robert Belle Burke (New York: Russell and Russell, 1962), vol. 1, p. 261.

Viète's innovations: See Jacob Klein's seminal work, *Greek Mathematical Thought and the Origin of Algebra,* tr. Eva Brann (New York: Dover, 1992), pp. 161–185; I cite Viète's *Introduction to the Analytical Art* from the translation in this volume by J. Winfree Smith (pp. 315–353). There is no book-length study of Viète; see my "Secrets, Symbols, and Systems," "François Viète, Father of Modern Cryptanalysis" and Helena M. Pycior, *Symbols, Impossible Numbers, and Geometric Entanglements* (Cambridge: Cambridge University Press, 1997), pp. 27–39.

Galileo on the Book of Nature: See "The Assayer" in *Discoveries and Opinions of Galileo,* tr. Stillman Drake (New York: Doubleday, 1957), pp. 237–238.

Descartes: See *The Philosophical Writings of Descartes,* tr. J. Cottingham, R. Stoothoff, and D. Murdoch (Cambridge: Cambridge University Press, 1985), vol. 1, pp. 2–3 (masks), 35, 290 (ciphers), 17, 19 (algebra), 20 (his method as "thread of Theseus"), 115 (abandons the study of letters), 141–142 (reaction to Galileo's condemnation), 142–143 (lordship over nature), 145 (his battles). For Descartes's stance toward his audience see Neil M. Ribe, "Cartesian Optics and the Mastery of Nature," *Isis* 88, 42–61 (1997).

Chapter 8: Kepler at the Bridge

Works by Johannes Kepler: *Mysterium Cosmographicum (The Secret of the Universe),* tr. A. M. Duncan (New York: Abaris Books, 1981), cited below as MC; *New Astronomy,* tr. William H. Donahue (Cambridge: Cambridge University Press, 1992), cited as NA; *The Six-Cornered Snowflake,* tr. C. Hardie (Oxford: Oxford University Press, 1966), cited as S; *The Harmony of the World,* tr. E. J. Aiton, A. M. Duncan, and J. V. Field (Philadelphia: American Philosophical Society, 1997), cited as HW.

References on Kepler: Max Caspar, *Kepler* (New York: Dover, 1993), cited as K; Arthur Koestler, *The Sleepwalkers* (New York: Viking Penguin, 1990); Gerald Holton, "Johannes Kepler's Universe" in *Thematic Origins,* pp. 53–74; D. P. Walker, "Kepler's Celestial Music" in his *Studies in Musical Science in the Late Renaissance* (Leiden: E. J. Brill, 1978), pp. 34–62; Owen Gingerich, *The Eye of Heaven: Ptolemy, Copernicus, Kepler* (New York: American Institute of Physics, 1993); Curtis Wilson, *Astronomy from Kepler to Newton: Historical Studies* (London: Variorum Reprints, 1989); *Planetary Astronomy from the Renaissance to the Rise of Astrophysics, Part A: Tycho Brahe to Newton,* ed. René Taton and Curtis Wilson, vol. 2A of *The General History of Astronomy* (Cambridge: Cambridge University Press, 1989); J. V. Field, *Kepler's Geometric Cosmology* (Chicago: University of Chicago Press, 1988); Bruce Stephenson, *Kepler's Physical Astronomy* (Princeton: Princeton University Press, 1987) and *The Music of the Heavens: Kepler's Harmony Astronomy* (Princeton: Princeton University Press, 1994); François De Gandt, *Force and Geometry in Newton's* Principia, tr. Curtis Wilson (Princeton: Princeton University Press, 1995), pp. 61–85; William Donahue, *Kepler's New Astronomy: The Central Argument* (Santa Fe, NM: Green Lion Press, forthcoming).

Kepler and the new mathematics: See J. V. Field, "The relation between geometry and algebra" in *Girolamo Cardano: Philosoph, Naturforscher, Arzt,* ed. E. Kessler (Wiesbaden: Harrasowitz Verlag, 1994), pp. 219–242 and my essay on "Kepler's Critique of Algebra," *Mathematical Intelligencer* 22 (2000).

Kepler and Fludd: See W. Pauli, "The Influence of Archetypal Ideas on Kepler's Theories" in C. G. Jung and W. Pauli, *The Interpretation of Nature and the Psyche* (New York: Pantheon Books, 1955), p. 196; corrected and supplemented by Robert S. Westman, "Nature, art, and psyche: Jung, Pauli, and the Kepler-Fludd polemic," in *Occult and scientific mentalities in the Renaissance,* ed. Brian Vickers (Cambridge: Cambridge University Press, 1984), pp. 177–229.

Accuracy of Ptolemy: See Curtis Wilson, "The Inner Planets and the Keplerian Revolution" in *Astronomy from Kepler to Newton.*

The Titius-Bode Law: See Stephenson, *Music of the Heavens,* pp. 10–11 and M. Ovenden, "Bode's Law—Truth or Consequences" in *Vistas in Astronomy,* ed. A. and P. Beer (Oxford: Oxford University Press, 1975), vol. 18, pp. 473–496.

Kepler and astrology: Kepler was not a traditional or superstitious practitioner; his astrological study disclaims the vulgar prediction of events, and envisages a much more complex interplay of celestial influences with earthly conditions; see Edward Rosen, "Kepler's attitude toward astrology and mysticism" in *Occult and scientific mentalities,* pp. 253–272.

Kepler and rhetoric: See N. Jardine, *The Birth of History and Philosophy of Science: Kepler's Defense of Tycho against Ursus* (Cambridge: Cambridge University Press, 1984), pp. 74–79. Stephenson emphasizes Kepler's rhetorical choice to present himself as groping; *Kepler's Physical Astronomy,* p. 3.

Kepler as "theory-laden": This is emphasized by Curtis Wilson, "Kepler's Derivation of the Elliptical Path," *Isis* 59, 5–25 (1968); "How Did Kepler Discover His First Two Laws?" *Scientific American* 226:3, 93–106 (1972), both included in his *Astronomy from Kepler to Newton.*

Kepler on Gilbert: See *Kepler's* Somnium, tr. Edward Rosen (Madison: University of Wisconsin Press, 1967), p. 100.

Kepler on "species": See William Donahue's helpful remarks on this crucial term, pp. 23–24 of NA. Kepler calls species "a variety of surface," as if it were an effluvium given off by the rotating sun, but the species is not a body independent of its source.

Kepler and the inverse-square law: See Curtis Wilson's "Newton and Some Philosophers on Kepler's Laws" and "From Kepler's Laws, So-called, to Universal Gravitation" in his *Astronomy from Kepler to Newton.*

Kepler's religious thought: See Jürgen Hübner, *Die Theologie Johannes Keplers zwischen Orthodoxie und Naturwissenschaft* (Tübingen: J. C. B. Mohr, 1975) and E. W. Gerde, "Johannes Kepler as Theologian," *Vistas in Astronomy,* 18, 339–367 (1975).

Kepler's sexual analysis of music: See HW, p. 242, 446 and Walker, "Kepler's Celestial Music," pp. 53–57; I hope to address the matter in *Beyond Passion: The Dialogue between Ancient and Modern Music* (in preparation).

Quotations from Kepler in text (NA unless specified): K 381 (epigram); S 7 (New Year's gift); S 33 (nothing occurs without reason); S 39 (almost Nothing); S 41 (flat); S 43 (formative faculty of Earth); Walker, "Kepler's Celestial Music," p. 245 (Geometrical Kabbalah); Pauli, "Influence of Archetypal Ideas," pp. 196–197 (I hold the tail); K 292 (I hate cabalists); MC 69 (Platonic solids); MC 61 (whole scheme); MC 19 (theologian); MC 51 (secret); MC 49 (God's motive); MC 53 (temple); MC 59 (Brahe); 30 (captive); 32 (untrackable star; Pliny; *Natural Histories* 2.17); K 56–57 (Kepler as teacher of rhetoric); 78 (Ptolemy); 48 (equivalence); K 103, 120–121 (public disclosure of secrets); 185 (Mars); 78 (Columbus); 79 (arguments); 46 (thread); 70–77 (synopsis); 119 (pretzel); 391 (magnet); 381–382 (species); 256 (loathing); 286 (Brahe); 86 (search); 451 (Christ!); 453 (sausage); 455 (blind from desire); 494 (errors lead to truth); 458, 508 (triumph); 543 (new light); 576 (O ridiculous me); 573 (Galatea, Virgil, *Ecologues* 3.64: "Galatea seeks me mischievously [*malo*], the lusty wench: / She flees to the willows, but hopes I'll see her first." *Malo* could also mean "with apples"); 492 (meandering route); 79 (pleasure); K 380 (geometry); K 380 (God's thoughts); K 375 (more lies beyond); MC 223 (never return to the same starting point).

Chapter 9: Newton on the Beach

Works of Newton: *The* Principia: *Mathematical Principles of Natural Philosophy,* tr. I. Bernard Cohen and Anne Whitman (Berkeley: University of California Press, 1999), cited as P; *The Mathematical Papers of Isaac Newton,* ed. D. T. Whiteside (Cambridge: Cambridge University Press, 1967–1980), cited as MP.

References on Newton: Alexandre Koyré, *Newtonian Studies* (Chicago: University of Chicago Press, 1968); Richard S. Westfall, *Never at Rest: A Biography of Isaac Newton* (Cambridge: Cambridge University Press, 1987), cited as NAR; Frank Manuel, *A Portrait of Isaac Newton* (New York: New Republic Books, 1979); Robert S. Bart, *Notes to Accompany the Reading of Newton's "Principia Mathematica"* (Annapolis, MD: St. John's College Press, 1968) [second edition]; Curtis Wilson, "Newton's Path to the *Principia*" in *The Great Ideas*

Today 1985, ed. Mortimer J. Adler (Chicago: Encyclopaedia Britannica, 1985), pp. 178–229 and "The Newtonian achievement in astronomy" in Taton and Wilson, *Planetary astronomy,* pp. 233–274; Dana Densmore, *Newton's* Principia: *The Central Argument* (Santa Fe, NM: Green Lion Press, 1995); De Gandt, *Force and Geometry.*

Comments of Halley and Laplace: see P 380 and Koyré, *Newtonian Studies,* p. 18; for Saint-Simon see Frank Manuel, *The Religion of Isaac Newton* (Oxford: Oxford University Press, 1974), p. 53.

Wonder and curiosity: See the superb discussion in Daston and Park, *Wonders and the Order of Nature,* pp. 303–350, which emphasizes the change from gawking wonder to scientific curiosity.

Voltaire: *The Elements of Sir Isaac Newton's Philosophy,* tr. John Hanna (London: Frank Cass and Company, 1967 [reprint of 1738 edition]), p. 4.

Pythagoras's lyre: Newton's niece Catharine Barton noted that "Sir Isaac used to say he believed Pythagoras' Musick of the Spheres was gravity, & that as he makes the sounds & notes depend on the size of the strings, as gravity depends on the density of matter" (NAR 510, n. 136). See J. E. McGuire and P. M. Rattansi, "Newton and the 'Pipes of Pan'," *Notes and Records of the Royal Society,* 21, 108–143 (1966).

Newton's alchemical work: See B. J. T. Dobbs, *The Foundations of Newton's Alchemy* (Cambridge: Cambridge University Press, 1983) and *The Janus faces of genius: The role of alchemy in Newton's thought* (Cambridge: Cambridge University Press, 1991); Richard Westfall, "The Role of Alchemy in Newton's Career" in Bonelli and Shea, *Reason, Experiment, and Mysticism,* pp. 189–232; J. E. McGuire, "Force, Active Principles, and Newton's Invisible Realm," *Ambix* 15, 154–208 (1968).

Newton's prophetic and historical materials: *Sir Isaac Newton's Theological Manuscripts,* ed. H. McLachlan (Liverpool: Liverpool University Press, 1950). Frank Manuel notes that Newton's "rules for interpreting the language of prophecy were a replica of those he insisted upon for interpreting the Book of Nature" (*The Religion of Isaac Newton,* p. 98). See also Frank Manuel, *Isaac Newton, Historian* (Cambridge: Harvard University Press, 1963), p. 28; Sarah Hutton, "More, Newton and the Language of Biblical Prophecy," in *The Books of Nature and Scripture,* ed. James E. Force and Richard H.

Popkin (Dordrecht: Kluwer Academic, 1994), pp. 39–53 (esp. pp. 46–49 on what Hutton calls "some kind of divine algebra" in the prophecies); Matania Z. Kochavi, "One Prophet Interprets Another: Sir Isaac Newton and Daniel" in *The Books of Nature,* pp. 105–122; Kenneth J. Knoespel, "Newton in the School of Time," *The Eighteenth Century* 30, 19–41 (1989).

Newton's use of ciphers: See NAR 243, 265, 403, 426, 515; Newton used anagrams for encipherment, including the provocative anagram for *Isaacus Neuutonus* read as *Iehova sanctus unus* ("Jehovah the holy one"). See also J. E. Hofmann, *Leibniz in Paris, 1672–1676* (Cambridge: Cambridge University Press, 1974), pp. 74–75; A. Rupert Hall, *Philosophers at War: The Quarrel between Newton and Leibniz* (Cambridge: Cambridge University Press, 1980), pp. 1–9. On enciphered discourse in Boyle's circle, see L. M. Principe, "Robert Boyle's Alchemical Secrecy: Codes, Ciphers and Concealments," *Ambix* 39:2, 63–74 (1992). For John Maynard Keynes on Newton as a "magician," see his *Essays in Biography* (New York: W. W. Norton, 1963), pp. 313–314.

Newton's "microscope": P, Book I, Proposition 6 (453–455). Thomas K. Simpson has called this figure the "hieroglyph for force." While connecting Newton's geometry helpfully with the project of decryption, the static term "hieroglyph" needs to be extended to encompass the dynamic aspect of his construction. See Thomas K. Simpson, "Science as Mystery: A Speculative Reading of Newton's *Principia*" in *The Great Ideas Today 1992*, ed. Mortimer J. Adler (Chicago: Encyclopaedia Britannica, 1992), pp. 96–109 at 140–144.

Newton uniting space and time through geometry: See De Gandt, *Force and Geometry,* pp. 12–32 and my paper on "Newton and hidden symmetry," *European Journal of Physics* 19, 151–153 (1998). For Galileo's separation of space and time, see his *Two New Sciences,* tr. Stillman Drake (Madison: University of Wisconsin Press, 1974), pp. xxii–xxiv.

Newton's use of algebra and geometry: See I. B. Cohen, "The Newtonian Revolution in Science," pp. 78–87 and Michael S. Mahoney, "The Mathematical Realm of Nature" in *The Cambridge History of Seventeenth-Century Philosophy,* vol. 1, pp. 702–755 and his "Algebraic vs. Geometric Techniques in Newton's Determination of Planetary Orbits," pp. 183–205 in *Action and Reaction,* ed. Paul Theerman and Adele F. Seeff (Newark: University of Delaware Press, 1993); De

Gandt, *Force and Geometry,* pp. 159–264; Pycior, *Symbols,* pp. 167–208.

Problems with infinity: For example, Newton expressed $1/(1 + x)$ as $1 - x + x^2 - x^3 + \ldots$ If $x = 1$, then $1 - 1 + 1 - 1 + \ldots = \frac{1}{2}$ (the average of successive sums), yet successive terms add to zero. An infinite series must be shown to be convergent, or self-contradictory consequences may follow. Also, George Berkeley raised questions about infinitesimals, "neither finite quantities, nor quantities infinitesimally small," in "The Analyst, or A Discourse Addressed to an Infidel Mathematician" (1734) in *A Source Book in Mathematics,* pp. 333–338, discussed in Pycior, *Symbols,* pp. 209–241.

The transcendence of e and pi: See Felix Klein, *Famous Problems of Elementary Geometry* (New York: Chelsea Publishing, 1962), pp. 49–77; for the historical background, see Michael S. Mahoney, "Infinitesimals and transcendent relations: The mathematics of motion in the seventeenth century," *Action and Reaction,* pp. 461–491 and my book *The Transcendent Scandal* (in preparation).

Newton's proof of transcendentality of oval areas: See the discussion of Newton's Lemma XXVIII of Book I (P 511–513) in V. I. Arnol'd, *Huygens and Barrow, Newton and Hooke* (Basel: Birkhäuser Verlag, 1990), pp. 101–105 and my "Newton and hidden symmetry."

Hypotheses non fingo: Newton himself explicitly invoked hypotheses in his physical theories, both astronomical and optical; see Koyré, *Newtonian Studies,* pp. 25–52.

Hidden core of nature: David Hume remarked that "while Newton seemed to draw off the veil from some of the mysteries of nature, he showed at the same time the imperfections of this mechanical philosophy; and thereby restored her ultimate secrets to that obscurity in which they ever did and ever will remain." *The History of England,* vol. 8, p. 332 (1773).

Newton on the beach: For biographical connections see NAR 863 and Frank Manuel, *Portrait,* pp. 388–389. Holton comments on the terror and sense of inadequacy that Newton must have felt in *The Scientific Imagination* (Cambridge: Cambridge University Press, 1978), pp. 268–274. For the "oceanic," see Freud, *Civilization and Its Discontents,* pp. 11 ff.

Circularity of concept of force: See G. W. F. Hegel's critique of the concept of force in *Phenomenology of Spirit,* tr. A. V. Miller (Oxford:

Oxford University Press, 1977), pp. 79–103. See also my essay "The Fields of Light," *St. John's Review* 28:3, 1–16 (1988–1989).

Newton's concept of force: Richard S. Westfall, *Force in Newton's Physics* (New York: American Elsevier, 1971), p. 450 ff discusses the interplay between passivity and activity in Newton's work. See also De Gandt, *Force and Geometry.*

Newton and comets: Thomas Simpson has suggested that Newton recognized in comets "a form of astronomical alchemy" since their tails undergo immense heating ("Science as Mystery," pp. 161–167). Newton suspects that "that spirit which is the smallest but most subtle and most excellent part of our air, and which is required for the life of all things, comes chiefly from comets" (P 926). As Simpson notes, "The heavens have become his laboratory." See also Simon Schaffer, "Comets and Idols: Newton's Cosmology and Political Theology" in *Action and Reaction,* pp. 206–231; Dobbs discusses the matter in *Janus face of genius,* pp. 230–243.

Experimentum crucis: The phrase was first used by Robert Boyle in 1662, and refers to Bacon's *instantia crucis* (Bacon 1.294, 4.180); see Brian Vickers, "Francis Bacon and the Progress of Knowledge," *Journal of the History of Ideas* 53, 495–518 (at 511, n. 45) (1992) and J. A. Lohne, "Experimentum Crucis," *Notes and Records of the Royal Society,* 23, 169–199 (1968). For Newton's use of Baconian "exclusions," see A. I. Sabra, *Theories of Light from Descartes to Newton* (Cambridge: Cambridge University Press, 1981), pp. 175–184; for contemporary controversies, see Dennis L. Sepper, *Goethe contra Newton* (Cambridge: Cambridge University Press, 1988); Charles Bazerman, *Shaping Written Knowledge* (Madison: University of Wisconsin Press, 1988), pp. 80–127; Stuart Peterfreund, "Saving the Phenomena or Saving the Hexameron? Mosaic Self-Presentation in Newtonian Optics," *The Eighteenth Century* 32, 139–165 (1991); Peter Dear, *Discipline and Experience,* pp. 232–243.

Citations from Newton (NAR unless otherwise noted): 863 (epigram); P 944 (electric spirit); P 382 (discover forces from the phenomena); 459 (smatterers): MP vol. 4, p. 277 (NAR 379; against Descartes); 510, n. 136 (Pythagoras); 309 (secrets of alchemy); Manuel, *Religion,* pp. 116, 108–109 (rules for interpretation; trial); 357 (purge the feces); McLachlan, *Theological Manuscripts,* p. 120 (hieroglyphic language); 326 (metaphor of watch); 346–348 (structure of the temple); P 940–941 (Pantokrator); P 453–455 ("microscope");

P 441–443 (evanescent); P 943 (our sea); 361 (Elaboratory); "The Key" (Keynes MS 18), in Dobbs, *Foundations of Newton's Alchemy,* pp. 251–255 (antimony); Dobbs, *Foundations of Newton's Alchemy,* p. 171 (Saturn swallowing his children); NAR 299 (nature as life); 299–301 (magnets); P 646 (genitum and flux).

Chapter 10: Einstein in the Boat

Writings by Einstein: *The Principle of Relativity* (New York: Dover Publications, 1923), cited as PR; *Essays in Science* (New York: Philosophical Library, 1934); *Ideas and Opinions* (New York: Random House, 1988); *Relativity: The Special and General Theory* (New York: Crown Publishers, 1961); *Sidelights on Relativity* (New York: Dover, 1983).

References on Einstein: Banesh Hoffmann and Helen Dukas, *Albert Einstein, Creator and Rebel* (New York: New American Library, 1972); Gerald Holton, *Thematic Origins,* pp. 191–398, and "Einstein and the Cultural Roots of Modern Science," *Daedalus* 127:1, 1–44 (1998); Ronald W. Clark, *Einstein: The Life and Times* (New York: World Publishing, 1971); Abraham Pais, *'Subtle is the Lord . . .': The Science and the Life of Albert Einstein* (Oxford: Oxford University Press, 1982) and *Einstein Lived Here* (Oxford: Oxford University Press, 1994).

Epigraph: Letter to Maurice Solovine, 28 March 1949, cited in Hoffmann and Dukas, *Albert Einstein,* p. 257.

Einstein in the boat: Chaim Tschernowitz, "A Day with Albert Einstein," *Jewish Sentinel* 1 (1931), cited in Clark, *Einstein,* p. 415.

Einstein and marriage: Pais, *'Subtle is the Lord . . .',* pp. 300–302. Quote about Spinoza: Friedrich Schleiermacher, *On Religion: Speeches to its Cultured Despisers,* tr. John Oman (New York: Harper & Row, 1958), p. 40. Einstein's apartness: Hoffmann and Dukas, *Albert Einstein,* p. 253.

Einstein, "Autobiographical Notes": Schilpp, *Albert Einstein,* pp. 9–11.

Relativity: Einstein's 1905 paper can be found in PR pp. 48–50. Hegel on "picture-thinking": see *The Phenomenology of Spirit,* pp. 119–122.

Minkowski: See PR, pp. 75–91; Edwin F. Taylor and John A. Wheeler, *Spacetime Physics* (San Francisco: W. H. Freeman, 1992) [second edition]; Peter L. Galison, "Minkowski's Space-Time: From Visual Thinking to the Absolute World," *Historical Studies in the Physical Sciences* 10, 85–121 (1979).

The world as four-dimensional: Einstein, *Relativity,* p. 55 and *Sidelights,* pp. 50–56.

Beyond the "merely-personal": See Einstein's "Autobiographical Notes" in Schilpp, *Albert Einstein,* vol. 1, p. 5 and Holton, *The Scientific Imagination.* For Kepler, see K 349, 387.

The temple of science: See "Principles of Research" in Einstein, *Essays in Science,* pp. 1–5.

Cosmic religious feeling: Einstein, *Ideas and Opinions,* pp. 38, 49 and Max Jammer, *Einstein and Religion* (Princeton: Princeton University Press, 1999).

Max Planck: *Scientific Autobiography* (New York: Philosophical Library, 1949), pp. 34–35; *Eight Lectures in Theoretical Physics,* ed. Peter Pesic (New York: Dover, 1998), p. 7.

General principle of relativity: PR, p. 113.

Dirac on quantum theory: P. A. M. Dirac, *The Principles of Quantum Mechanics* (Oxford: Oxford University Press, 1958 [fourth edition]), p. vii.

Einstein and Spinoza: See Peter Bucky, *The Private Albert Einstein* (Kansas City: Andrews and McMeel, 1992), pp. 112–113; Hoffmann and Dukas, *Einstein,* p. 94.

"Although he lived 300 years before . . .": Einstein's introduction to R. Kayser, *Spinoza, portrait of a cultural hero* (New York: Philosophical Library, 1946), p. xi, quoted in Pais, *Einstein Lived Here,* p. 130, who also comments that "it could well be that he was thinking about himself" in the passage cited.

"Nothing in the universe is contingent": *The Collected Works of Spinoza,* tr. Edwin Curley (Princeton: Princeton University Press, 1985), vol. I, p. 433.

"I believe in Spinoza's God": Einstein's response to the question "Do you believe in God?", addressed to him by Rabbi Herbert Goldstein in 1929; see my paper "Einstein and Spinoza: Determinism and

Identicality Reconsidered," *Studia Spinozana* 12, 193–201 (1996). See also Jammer, *Einstein and Religion.*

The Bohr/Einstein dialogue: See Niels Bohr, "Discussion with Einstein on Epistemological Problems in Atomic Physics" in Schilpp, *Albert Einstein,* vol. 1, pp. 201–241, along with Einstein's "Reply to Criticisms" in vol. 2, pp. 665–676; see also David Kaiser, "Bringing the human actors back on stage: the personal context of the Einstein-Bohr debate," *British Journal of History of Science* 27, 129–152 (1994).

Identity in quantum theory: See my papers on "The principle of identicality and the foundations of quantum theory I, II," *American Journal of Physics* 59 (1991), 971–978 and my book *Beyond Identity: Physics, Philosophy, Literature* (in preparation).

The two levels of quantum theory: See Abraham Pais, *Inward Bound* (Oxford: Oxford University Press, 1986), pp. 255–261.

Egyptian labyrinth: Herodotus, *History* 2.148.

Polyalphabetics and the one-time system: See David Kahn, *Codebreakers,* pp. 398–410, 979–984; on Enigma machines and the "Magic" decryptions, see *Codebreakers,* pp. 394–512. The Einstein-Podolsky-Rosen experiment can be used to transmit that random key to distant places; see Charles H. Bennett, Gilles Brassard, and Artur K. Ekert, "Quantum Cryptography," *Scientific American* 263:10, 50–57 (1992).

Aristotle's joke: "To prove what is obvious by what is not is the mark of a man who is unable to distinguish what is self-evident from what is not." *Physics* Book II, Chapter I 193a5–7.

The empty maze: Holton evokes "the labyrinth with the empty center, where the investigator meets only his own shadow and his blackboard with his own chalk marks on it, his own solutions to his own problems" in *Thematic Origins,* pp. 30–31, 36 [only in first edition].

Acknowledgments

My thinking has been greatly helped by the liberating influence of St. John's College, whose program lifted disciplinary barriers and stimulated me to study great works freshly. I remember my fellow students with gratitude.

I salute Larry Cohen and his associates at the MIT Press, who had the vision and courage to publish this book.

Millicent Dillon helped me find the courage to know my mind and to speak it. Curtis Wilson generously shared his deep knowledge. The brilliant work of John C. Briggs on Bacon deeply influenced me, as did that of Nieves Mathews. I thank them for their advice and encouragement.

Finally, Ssu, Andrei, and Alexei—more than the world.

Index